IT BEGAN WITH A STONE

IT BEGAN WITH A STONE

A History of Geology from The Stone Age to the Age of Plate Tectonics

HENRY FAUL

CAROL FAUL

Department of Geology
University of Pennsylvania

A Wiley-Interscience Publication

John Wiley & Sons

New York • Chichester • Brisbane • Toronto • Singapore

Library of Congress Cataloging in Publication Data:

Faul, Henry.
It began with a stone.

"A Wiley-Interscience publication."
Bibliography: p.
Includes index.
1. Geology—History. I. Faul, Carol. II. Title.

QE11.F38 550'.9 83-3683
ISBN 0-471-89735-3
ISBN 0-471-89605-5 (pbk.)

Printed in the United States of America

10 9 8 7 6 5 4 3 2 1

To the Faculty, Staff,
and Graduate Students of the
Penn Department of Geology

Preface

Henry Faul died before the manuscript of *It Began with a Stone* was completed. Virtually all the text was organized—mostly I have filled in gaps, completed the references, and chosen some illustrations.

Henry was writing this book because after 40 years as a professional geologist, he felt he was beginning to understand the people and the things they wrote which shaped the development of the Earth sciences. It is clearly his view of the history of his science. He had made a few notes of what he wanted to tell the readers of his book:

> *Some historians think they are able to write about their long departed chosen subjects from the point of view of contemporary witnesses, in the spirit of time long past. I know I have no such powers, and my view is perforce from the present, based on my position and experience in this world.*

It Began with a Stone is a comprehensive look at the history of geology. It describes people who were curious—who and what gave them ideas, why and how they transmitted those ideas. To emphasize the progression of ideas he chose specific people who influenced the directions in which geology was going (although more than one of those directions led to an abyss and some worthy minds tumbled in).

Chapters cover broad periods of time and are organized more or less by the countries of the participants. *It Began with a Stone* differs from other histories of geology in its scope and in its approach. Three books, especially, written by geologists and describing the development of the Earth sciences are valuable predecessors. Karl von Zittel's *Geschichte der Geologie und Paläontologie*, published in 1899 (English edition: *History of Geology and Paleontology*, 1901) is an encyclopedic survey with emphasis on Continental European developments in the nineteenth century. In a different approach, Archibald Geikie wrote about the most important personalities but little on the lesser lights in *The Founders of Geology* in 1905. Both were a bit chauvinistic and didn't look hard across the Atlantic. Canadian geologist Frank Dawson Adams in 1938 wrote *The Birth and Development of the Geological Sciences*, a scholarly and orderly review of mostly early geologic development through the Classical period.

None of these works has the wide perspective of this book and none includes a list of references. Henry decided that an extensive bibliography was important—there are over 550 entries of works mentioned in the text. He felt that the people who developed the practice of geology in France had been overshadowed by the highly visible and eloquent British and that the chapters on French geology were among the most important contributions of his book.

The style is Henry's—facts and opinions, thoroughly documented. The facts he started to collect early. As a student he learned paleontology from Bedřich Bouček and collected fossils where Barrande did, from the outcrops around his native Prague. He worked in a Swedish mine for a summer, went to MIT for a degree in mining, then to Michigan State where he ended up studying rugose corals and their growth lines. One day, as he told it, he fixed a broken Geiger counter and soon found himself recruited by the Manhattan Project. Henry learned to like geophysics; that became his PhD subject at MIT and the focus of his work at the U.S. Geological Survey. He was proud to be a "plumber"—he designed and constructed instruments including mass spectrometers for the U.S. Government and for many universities. He was a plumber who used his tools for solving problems, especially in geochronology, such as determining temporal relationships among crystalline-rock masses in New England, reconstructing the chronological development of Alpine structures, and working out the age relationships of the intrusive ring complexes in eastern Egypt. Besides numerous publications on geology, he was editor of *Nuclear Geology* (1954: Wiley), a classic compilation of papers on nuclear geophysics. (He was perhaps more prolific as a writer of books and articles on antique furniture—under a pseudonym.)

Henry came to the University of Pennsylvania for the stimulation provided by students and to Philadelphia because so much of importance in the Earth sciences began here, and because that's where the books were. Henry and I collected many books that were germinal to geology and he lovingly repaired them because they were so essential a part of his science. Books are an important part of this book. Henry began to write on the history of geology to review current writings about which he had strong opinions. He decided he needed to know more to document those opinions. He traveled widely—North America, Europe, the Soviet Union, North Africa, the Middle East. He used the eight or more languages he spoke and read to ask questions and to read the sources he would write about.

At Penn, Henry taught a course on the history of geology and learned a lot from the students. A research project by an Algerian, for example, got him interested in early Arab geology. Because there was no book suitable for a course text, he gathered together his early geology books, his notes, and photographs, and he wrote. The outline for *It Began with a Stone* was organized in 1974 in Shiraz, Iran, were we taught during a sabbatical leave. He worked on it for the rest of his life.

This book is for people who care about the Earth and how the scientific study of it developed. As a college text it is suitable for seniors and graduate students. I hope that professional geologists and historians of science will like the scope of the work—as a scholarly contribution to the history of geology and as an interesting book through which to browse.

Many people helped Henry and then me. Claude C. Albritton, Jr. was of great help to us both, academically and as an encouraging friend. Claude, Clifford J. Awald, Robert F. Giegengack, Jr., Robert Hazen, Joan Heubusch, and Hermann W. Pfefferkorn all read the entire manuscript and discussed—and argued—with us. Hermann, in addition, helped complete the photography. Victor Coutant, Joan Eyles, Rhoda Rappaport, and Charles Rosenberg read sections and provided ideas. Pauline Wong photographed many of the illustrations from books, very ably, as you can see. (Photographs of other than book illustrations are by Henry Faul unless otherwise credited.) Yoshikazu Ohashi advised us on computer techniques. Penn graduate students Paula Luborsky and Robb Turner convinced the word processor to complete the manuscript and Paula typed in the final corrections. The reference librarians of Penn's Van Pelt Library assisted with many esoteric requests. Wiley editor Everett Smethurst was so helpful and encouraging, he became a good friend. If I have omitted mentioning anyone who helped Henry, it is because I am not aware of their contribution. Please forgive me.

Readers will note that there is a paucity of women mentioned in this history of the Earth sciences. I wish to acknowledge the contributions of the multitude of anonymous women who were surely there influencing the "founders" of geology.

Any errors are my responsibility, not even Henry's. He would have been appalled that any had slipped into his book.

CAROL FAUL

Philadelphia
Spring 1983

Contents

List of Illustrations

CHAPTER ONE

Antiquity

VESTIGES OF A BEGINNING

Geology began when early man first picked up a stone, considered its qualities, and decided that it was better than the stone he already had. Good stones were useful and they were collected, mined, and traded. Knowing which stone was good and where it could be found may not have been geology in any present sense, but compared with the level of other arts and sciences at the time it was advanced expertise.

It does not matter what he intended to do with the stone. Whether a tool or a weapon, a better stone was an advantage in the competition for food and territory. Without thinking about it, the ancestor was improving his chances by finding better stones. From the very beginning, geology was a factor in human survival.

Practical knowledge of water, stone, clay, sand, and (much later) metal expanded as cultures developed. The testimony of ancient stone implements, pottery, mineral pigments, dimension stone, metal castings, and glassware proves that early man developed complex industries and sufficient geological knowledge to find and produce the raw materials required for those industries. Underground flint mines operated in Belgium (Rutot, 1913) and in southern England more than 5000 years ago (Smith, 1974). People of the Vinča culture, about 4500 years ago, mined copper in 20 meter shafts at Rudna Glava in the Bor mining district in Yugoslavia (Jovanović, 1971). Egyptian tomb paintings of 3500 years ago show foundries working and metal ingots being counted and carried. Three dozen copper ingots were found in the wreck of a Syrian ship that foundered on the rocks of Cape Gelidonya in southern Turkey 3200 years ago (Bass, 1967).

Written records are much rarer than the archaeological evidence. Cuneiform tablets from Mesopotamia make references to artificial gemstones as early as 4000 years ago, and later texts give recipes and procedures for making glass (Oppenheim, Brill, Barag, and Saldern, 1970). Elaborate glassware was manufactured and traded in both Mesopotamia and Egypt some 3500 years ago. Fragments of a papyrus roll of geological significance are preserved in Turin (Scamuzzi, 1965). It shows roads, a well, and "the mountains where gold is mined" and has been the subject of much discussion among Egyptologists and geologists alike (Hume, 1937). It was written about 1150 B.C. and appears to be a court record of a dispute about a statue that was ordered for the temple of Ramses II in Gurnah (the Ramesseum of Thebes). The statue was made of a special and presumably sacred stone, but was abandoned unfinished along the way to the Nile (Goyon, 1949).

The Turin papyrus is the oldest known map. It is not drawn to scale, but has north at the top. Goyon (1949) has identified the area of the map as present-day Wadi Hammamat, where gold is known to have been mined since pharaonic times, and where unusual green decorative stones (now described as metavolcanic) are known to have been quarried.

The literature of early technology is scarce because practitioners of the commercial arts rarely wrote anything. On the contrary, their interest was to keep the business in the family, and they could have been expected to do their best to conceal their techniques and their sources and to maintain trade secrets. When a profitable industry becomes too well known, it is bound to be taxed by whatever authority can reach it, and that is the basic reason why miners have always been reluctant to volunteer much information about the extent of their business. Much of the mystery that surrounded the sources of metals and minerals in antiquity was deliberately generated by the people in the trade. A little mystique never hurt any endeavor, and the book-writing philosophers could not be trusted. They had ties to the court and might be in league with the tax collector.

Fossils have intrigued man since early antiquity, and occasionally they appear in graves (Oakley, 1965), together with other unusual stones. The writings of ancient Greece included numerous mentions of fossil shells found in the hills high above the sea (Pease, 1942), and such discoveries started people thinking about great inundations. Large fossil bones found in alluvial strata led to stories of giants. Some fossils were thought to be endowed with mystic powers, judging by all the popular superstitions in the ancient and medieval literatures, some of which still survive.

Greater questions about man's environment led to discussion of geological subjects. Floods, earthquakes, and volcanic eruptions were recorded since earliest historic times and their significance was discussed. More abstruse questions of the origin of the world, the nature of life, and the provenance of man were always taken up in one context or another. Call it religion, or natural theology, or theoretical geology—the distinctions are not necessarily fixed, even now.

GREECE

Ruins of Greek construction abound in the Mediterranean region, but as a matter of perspective one must realize that all written records of ancient Greece are exceedingly scarce. It is difficult to estimate how much Greek literature has survived in any form, but it almost certainly must be less than 1 percent. Very few of the surviving works are in their original form, almost all are corrupted by careless copying over many centuries, and few contain much information that would permit dating them even indirectly. Nevertheless, it is clear that ancient Greece was the scene of much geological thinking (Schvarcz, 1868) and that many of the seminal ideas eventually found their way into printed books.

The fragmentary kaleidoscope of Greek science records isolated glimpses of important geological understanding. In the sixth century B.C., the Pythagoreans were the first to teach that the Earth was round because it cast a round shadow on the moon in eclipses. Eratosthenes (ca. 276–ca. 195 B.C.), the second director of the Museum Library in Alexandria, devised a method of measuring the diameter of the Earth's sphere. He observed that on the summer solstice, the longest day of the year, the sun stood at an angle of one-fiftieth of a circle from vertical in Alexandria, but was directly overhead in Syene (now Aswan). He had no accurate way of measuring the north-south distance between the two places (the length of the meridian), but he made a reasonable estimate and calculated a remarkably accurate value for the Earth's diameter. Later commentators have pointed out the obvious sources of error in his estimate, but that does not diminish the brilliant simplicity and fundamental grandeur of the experiment. There is some uncertainty about the units he used, but his result appears to be only about 20 percent larger than modern determinations.

Extending his calculations to include some doubtful sun-angle determinations and rough distance estimates, Eratosthenes then attempted to establish a coordinate grid for the whole ancient world. The result was a distinct improvement in the world map of the day, but the unreliability of the data was recognized and severely criticized by the astronomer Hipparchus of Bithynia (?–after 127 B.C.). In categorically rejecting conclusions based on inadequate data, Hipparchus was being very scientific, but in failing to concede that an inspired guess is better than no information at all he set geography back a fair distance. It was this same Hipparchus who pioneered the quantitative approach to astronomy and developed the precession of the equinoxes. Almost all of his original writings are lost and his work is known mostly from references to it in Ptolemy's *Almagest*, written about three centuries after Hipparchus died.

Herodotus of Halicarnassus (?484–?425 B.C.), the great historian of Greek antiquity, concerned himself mainly with political and military history, but he had traveled as far as the Black Sea, Mesopotamia, and Egypt and made accurate geological observations. He was aware that earthquakes cause large-scale fracturing and thus may shape the landscape (History, VII, 129). He noted the sediment carried by the Nile and estimated the amounts of deposition from the annual floods in the Nile Valley and the growth of the great delta. "Egypt . . . is an acquired country, the gift of the river," he wrote (Herodotus, 1858, II, 5), and throughout those discussions he displayed a remarkable understanding of the vastness of geologic time (Harrington, 1967).

ARISTOTLE AND THEOPHRASTUS

The natural history of Aristotle (384–322 B.C.) exercised an overwhelming influence for almost 2000 years and his followers held a virtual monopoly on the subject in the early Middle Ages. The work credited to him survives in a transcription made by Andronicus of Rhodes about 250 years after Aristotle's death. Original writings are here intermingled with drafts, notes, and interpretations, not in chronological order. The oldest surviving copies of parts of this text date from the ninth century, but much of it is known only from copies made in the eleventh and twelfth centuries. Fields that would now be known as ethics, logic, mathematics, physics, and biology are all one continuum in Aristotle's writing, and his work in the sciences (often maligned by later generations) must be considered within the frame of his enormous contribution to philosophy. To Aristotle, the purpose of science was little more than to illuminate philosophical principles.

He wrote little on geological subjects. The *Meteorologica* (Aristotle, 1952) mentions earthquakes and ascribes them to underground winds produced by the interaction of the two great fundamental forces: hot and cold. The division of all nature into four interchangeable elements (fire, air, water, earth) corresponding to the basic physical properties (hot, cold, wet, dry) of various materials including rocks and minerals was central in the natural philosophy of the Greeks, and it survived well into the Renaissance.

Legions of interpreters have endowed Aristotle's views with extraneous definition. He is often presented as advocating the spontaneous generation of insects, frogs, and fish. That was indeed the prevailing view in his time, and the idea remained until late into the seventeenth

FIGURE 1.1. Earthquake. (Apocalypsis Sancti Johannis, 1474: courtesy Lessing J. Rosenwald Collection, Library of Congress, Washington, DC)

century, but Aristotle himself had taken a cautious position insofar as we know his writings. He wrote nothing about fossil shells, but is widely credited with reporting that fish can grow in rocks. That misconception may have come through Seneca's *Questiones Naturales,* from a misreading of a minor work of Theophrastus (Theophrastus, 1866, fragment 171), or from a collection of hearsay reports erroneously attributed to Aristotle in antiquity (*On Marvellous Things Heard,* par. 72–74, see Aristotle, 1936, pp. 265–267) which actually may be only a summary of that same work by Theophrastus. The original texts are not quite clear, but they apparently refer not to fossil fish but to lungfish and other live fish presumably dug from the mud of dried-up rivers in Asia Minor (Pease, 1942).

Theophrastus of Eresus (ca. 371–ca. 287 B.C.) followed Aristotle in the Peripatetic school in Athens and was a major star on the intellectual scene for most of his long life. His writings are largely lost, but two major works in botany survive, plus several smaller studies and many fragments, including the short treatise *Peri Lithon (On Stones),* written about 314 B.C. That would make Theophrastus the author of the earliest known geology book, but the honor is misleading (for a well-annotated translation see Caley and Richards, 1956). *On Stones* has less than 5000 words and amounts to a catalog of the mineral substances found in the Athenian trade at the time. The introduction reflects the fundamental difficulties the Greeks encountered in categorizing natural substances in terms of the four properties they considered basic.

Theophrastus had access to the silver mines of Laurium, a state-run enterprise, and he found some reliable informants in the marketplace. The treatise includes several accurate descriptions of contemporary technology, such as the manufacture of white lead and verdigris for pigments, and the use of the touchstone for testing the purity of precious metals. Other statements show gaps of information, notably in the discussion of glass. By Theophrastus' time, glass had been a common article of trade in the eastern Mediterranean region for centuries, but he obviously could not find out how it was made. He records several superstitions, apparently aware that is what they were, and makes it clear that he is not necessarily endorsing them. He writes of stones that have the wonderful power, "if that be true," to give birth to young (the *aetites* or "eagle stones") and of *lyngourios* (Latin *lyngurium*), supposed to be the congealed urine of the lynx.

Some of Theophrastus' names for rocks and minerals have survived more or less in the meaning he used: alabaster, agate, amethyst, chrysocolla, cinnabar, crystal (quartz), and gypsum. Other terms can be translated from their context (with varying degrees of certainty): amber, azurite, lapis lazuli, lodestone (magnetite), malachite, obsidian, orpiment, and realgar. The meanings of the rest, including about half of all the terms he used to describe his stones, remain obscure. Some terms are used loosely (as indeed they still often are), such as alabaster to include both gypsum and calcite, and *gypsos*, to mean not only what we now call gypsum plaster, but also fuller's earth and slaked lime. The word *smaragdos* (emerald) describes several unrelated green stones, including malachite and chrysocolla. *Haimatitis, iaspis,* and *sappheiros,* to cite three conspicuous examples, are described as having properties inconsistent with what we now call hematite, jasper, or sapphire.

Peri Lithon is preserved in 13 manuscripts (Eichholz, in Theophrastus, 1956), but the oldest was copied about 1600 years after Theophrastus had written it, and all of the known texts are seriously corrupted. Many passages are unintelligible in their present forms, and others are open to different interpretations. Thus we may never be able to assay the full extent of what Theophrastus knew and did not know about mineralogy. The response to *Peri Lithon* was not great. Pliny quotes it frequently and, incidentally, makes light of the lynx story (*Natural History* XXXVII, 3). Lyngurium is amber, says Pliny with conviction, and it comes by trade from the shores of the Baltic. Beyond that there is little mention of Theophrastus by name until the Renaissance, when Geor-

FIGURE 1.2. Amber hunters in East Prussia. Some dig for it in the "blue earth," others fish for it with large nets in the roiling surf. (Hartmann, 1677)

gius Agricola uses him as classical authority to back up his own empirical observations.

THE GEOCENTRIC UNIVERSE

Considering man's preoccupation with himself as the center of things, it is perhaps not surprising that Greek astronomy, having evolved from ancient ritualistic observations of heavenly bodies, developed in a system with the Earth always at the center. Very little of the early Greek astronomical literature survives in any form, and the best view of the Greek system is obtained only in a relatively late work, the *Almagest* of Claudius Ptolomaeus (ca. 100–ca. 170). He wrote it in Alexandria while that city was still a center of learning, and it is preserved in its original Greek, in translation into Arabic, and in retranslation into Latin. It is a clear and practical textbook of astronomy, presenting the geometry of the motions of the "planets" recognized by the Greeks (moon, Mercury, Venus, sun, Mars, Jupiter, and Saturn—in the order of supposed distance from the Earth). The methods it describes and the data it gives in its tables are sufficient to calculate planetary positions within a few minutes of arc, or within the range of precision of the astronomical instruments available at the time. Even more important in the contemporary scale of priorities, the *Almagest* alone was enough to allow calculations for correctly predicting eclipses.

Viewing the Earth as the center of the solar system makes planets appear to be moving in complicated curves against the background of the "fixed" stars, and the mathematical description of these apparent motions is not trivial. Precluding as it did any possibility of understanding the physics of the situation, the geocentric approach was a scientific blind alley, perhaps the grandest of all time, but it posed a tremendous challenge to both theory and observation. As their measuring techniques advanced, the Greek astronomers and their Islamic successors tried everything in their power to describe and explain this ever-increasing complexity.

After the early estimates of Eratosthenes, more and better latitudes and longitudes were determined, and Ptolemy in the second century was able to collect a respectable number of them in his *Geographia*, written after the *Almagest*. Giving latitudes and longitudes of cities and well-known geographic features, the *Geographia* was, in effect, the first atlas. To calculate latitude, then as now, it was necessary only to measure the elevation of the sun at noon on a given date, but longitude presented a much more difficult problem. The only accurate method then available for determining the difference in longitude between two places required a comparison of the exact times of the same eclipse observed from those two places. Measuring time was difficult and very few such determinations had been made, so Ptolemy had to rely on the notoriously inaccurate distance estimates made by sailors and travelers in the east-west direction. Nevertheless, his *Geographia* remained a standard reference for some 1400 years.

ROME

Several writers of Graeco-Roman antiquity had great influence on the development of geology, not through original work of their own, but simply because their manuscripts happened to survive and ultimately reached the printing press. Lucretius (ca. 99–ca. 55 B.C.), Strabo (ca. 63 B.C.–after A.D. 21), Seneca (ca. 3 B.C.–A.D. 65), and Pliny the Elder (ca. 23–79) are the most prominent in that group. Their works are still the main sources of what little is known about ancient geology, and for centuries they have been copied, printed, read, and searched for ideas.

FIGURE 1.3. Roughed out and unfinished capital left behind in the Roman granite quarry of Mons Claudianus, west of Safaga, Egypt, between the Nile and the Red Sea. The penal quarries were active in the second century, A.D.

Democritus of Abdera (fifth century B.C.) is credited with the brilliant idea that all matter is composed of small indivisible particles called atoms. Actually, he may not have originated the concept but perhaps adapted it from Leucippus or from other sources now lost. Only small fragments of Democritus' work survive, most of them attributed to him unreliably, and none mentions the atomic theory. Later the idea was carried on by Epicurus (341–270 B.C.) and his followers, was discussed adversely by Aristotle and Theophrastus, and received its finest presentation only very much later in a splendid hexametric poem *De Rerum Natura* (*On the Nature of the Universe*) by Titus Lucretius Carus, who lived in Rome about the time of Cicero. The poem has survived in a single copy and practically nothing is known about the author.

Lucretius analyzed man's position in the world in the light of Epicurean philosophy. He renders his conclusions in such a magnificently quotable way that trying to summarize or paraphrase them would be sheer vanity.

Excerpts from De Rerum Natura

On Creation

Nothing can ever be created by divine power out of nothing. The reason why all mortals are so gripped by fear is that they see all sorts of things happening on the Earth and in the sky with no discernible cause, and these they attribute to the will of a god (Book I, 152).

The theory that the gods deliberately created the world in all its natural splendor for the sake of man, so that we ought to praise this piece of divine workmanship and believe it eternal and think it a sin to unsettle by violence the ever lasting abode established for mankind by the ancient purpose of the gods . . .—this theory with all its attendant fictions is sheer nonsense. For what benefit could immortal and blessed beings reap from our gratitude, that they should undertake any task on our behalf? (II, 190).

Nature is free and uncontrolled by proud masters and runs the universe by herself without the aid of gods (I, 419).

On the Universe

All nature as it is in itself consists of two things—bodies and the vacant space in which the bodies are situated and through which they move in different directions (I, 419).

In all dimensions alike, on this side or that, upward or downward through the universe, there is no end. . . . Granted, then, that empty space extends without limit in every direction and that innumerable seeds are rushing on countless courses through an unfathomable universe under the impulse of perpetual motion, it is in the highest degree unlikely that this earth and sky is the only one to have been created and that all those particles of matter outside are accomplishing nothing (II, 1047).

On Atoms and Their Combinations

Nature resolves everything into its component atoms and never reduces anything to nothing (I, 216).

The number of different forms of atoms is finite. If it were not so, some of the atoms would have to be of infinite magnitude (II, 481).

It must not be supposed that atoms of every sort can be linked in every variety of combination. If that were so, you would see monsters coming into being everywhere (II, 699).

On the Changing Landscape

Part of the soil is reconverted to flood water by the rains, and gnawing rivers nibble at their banks. And whatever Earth contributes to feed the growth of others is restored to it. It is an observed fact that the universal mother is also the common grave (V, 250).

The uprooted boulders rolling down a mountainside proclaim their weakness in the face of a lapse of time by no means infinite; for no sudden shock could dislodge them and set them falling if they had endured from everlasting, unbruised by all the assault and battery of time (V, 311).*

Lucretius' poetry has an interesting characteristic: In every intellectual environment since the Roman, people have easily read more into his poetry than actually was there. Lucretius presents the quintessence of Greek contributions to science and he accurately reflects an inspired philosophical attitude, but he is not reciting the results of scientific observations. His atoms and his universe are hypothetical concepts, and they were quite unacceptable to most of his contemporaries. In the modern sense he probably would have had much difficulty finding a publisher for his book.

Unwittingly, Lucretius illustrates the difficulties ancient Greek philosophers encountered in trying to come to terms with elementary mechanics. He relates the boundlessness of the universe in a language that relativistic astrophysicists could savor, but vainly struggles for an explanation of the forces that would hold up the Earth within it. Lucretius excites the imagination and that is why he was unloved by every church, gleefully espoused by the Enlightenment, and regarded as a prophet ever since.

Strabo traveled widely in the Mediterranean region and his work was mainly in history, but little of that survives. His later book on geography is preserved and it is the best source we have for the work of Eratosthenes and the critique of it by Hipparchus. Strabo's own critical judgment is flawed: He disparages Eratosthenes, ignores the direct observations of Herodotus, but takes Homer's mythical geography at face value. These soft spots are easily found, however, and they do not lessen the value of his information relating to geological subjects.

* Lucretius, *On the Nature of the Universe*, R. E. Latham trans. (Penguin Classics 1951) copyright © R. E. Latham 1951, pp. 31, 33, 39, 74, 80, 91, 92, 178, 180; reprinted by permission of Penguin Books, Ltd.

Lucius Annaeus Seneca, a philosopher, playwright, and politician at the court of Nero, wrote around the year 62, among many other works, one book of natural philosophy that has some geological interest, the *Quaestiones Naturales*. It discusses volcanoes and earthquakes, very much in the Aristotelian context, and mentions the possibility of fish living in rocks. The works of Seneca were widely read and the *Quaestiones* was an important vehicle for transmitting these Aristotelian and pseudo-Aristotelian ideas into western Europe.

By far the most important reporter of Roman science is Gaius Plinius Secundus, called Pliny the Elder. He had a long career as a public official under Vespasian and was Prefect of the Fleet in Misenum (10 miles west of Naples) when Vesuvius erupted for the first time in Roman history. He sailed across the bay to investigate, went ashore near Stabiae in a hail of pumice and ashes, and died there of what is usually described as asphyxiation but what may have been a heart attack. Most of what we know about his life comes from three letters written by his nephew and heir, Gaius Plinius Caecilius Secundus (?62–ca. 113), called Pliny the Younger or Pliny the Consul. It is not known when Pliny the Younger, who was only about 18 when his uncle died, wrote but presumably it must have been long after the events (*Epistolae*, III, letter 5, and VI, letters 16 & 20).

Excerpts from Two Letters from Pliny the Younger to His Friend, the Historian Cornelius Tacitus (VI.–16, 20)

My uncle was stationed at Misenum (now Punto di Miseno), in active command of the fleet. On August 24th, (79 A.D.), in the early afternoon, my mother called his attention to a cloud of unusual size and shape. He called for his shoes and climbed up to a place which would give him the best view of the phenomenon. It was not clear at that distance from which mountain the cloud was rising (it was afterwards known to be Vesuvius); its general appearance can best be described as being like an umbrella pine, for it rose to great heights on a sort of trunk and then split off into branches, I imagine because it was thrust upward by the first blast and then left unsupported as the pressure subsided, or else it was borne down by its own weight so that it spread out and gradually dispersed. In places it looked white, elsewhere blotched and dirty, according to the amount of soil and ashes carried with it.

He gave orders for the warships to be launched and went on board himself with the intention of bringing help to the people, for this lovely stretch of coast was thickly populated. He hurried to the place which everyone else was hastily leaving, steering his course straight for the danger zone. Without fear, he was describing each new movement and phase to be noted (by his secretary) as he observed them. Ashes were already falling, hotter and thicker as the ships drew near, followed by bits of pumice and blackened stones, charred and cracked by the flames; then suddenly they were in shallow water, and the shore was blocked by the debris from the mountain. For a moment my uncle wondered whether to turn back, and the helmsman advised it, but he refused, telling him that Fortune stood by the courageous and they must make for Pomponianus at

Stabiae. He was protected there by the breadth of the bay (for the shore gradually curves around a basin filled by the sea) so that he was not yet in danger, though it was clear that this would come nearer as it spread. Pomponianus had already put his belongings on board ship, intending to escape when the unfavorable wind abated. This wind was, of course, full in my uncle's favor, and he was able to bring his ship in.

Meanwhile, on Mount Vesuvius broad sheets of fire and leaping flames blazed at several points, their bright glare enhanced by the darkness of night. My uncle tried to allay the fears of his friends by repeating that these were only bonfires left by the peasants in their terror, or else empty houses on fire in the districts they had abandoned. Then he went to rest and certainly slept, for as he was a stout man, his breathing was loud and heavy and could be heard by the people coming and going outside his door. By this time the courtyard giving access to his room was full of ashes mixed with pumice-stones, and its level had been raised so that if he had stayed in the room any longer, he would not have been able to get out. He was awakened, came out, and joined Pomponianus and the rest of the household who had sat up all night. They debated whether to stay inside or take their chance in the open, for the buildings were now shaking with violent shocks, and seemed to be swaying to and fro as if they were torn from their foundations. Outside, on the other hand, there was the danger of falling pumice-stones, even though they were light and porous. After comparing the risks, they chose the latter. As a protection against falling objects they put pillows on their heads tied down with cloths.

Elsewhere there was daylight by this time, but they were still in darkness, blacker and denser than any ordinary night, which they relieved by lighting torches and lamps. My uncle decided to go down to the shore and investigate the possibility of escape by sea, but he found the waves still wild and dangerous. A sheet was spread on the ground for him to lie down and he repeatedly asked for cold water to drink. Then the flames and smell of sulphur which gave warning of the approaching fire drove the others to take flight and roused him to stand up. He stood leaning on two slaves and then suddenly collapsed, I imagine because the dense fumes choked his windpipe which was constitutionally weak and narrow and often inflamed. When daylight returned on the 26th—two days after the last day he had seen—his body was found intact and uninjured, still fully clothed and looking more like sleep than death.

Meanwhile, my mother and I were at Misenum . . . For several days past there had been earth tremors which were not particularly alarming because they are frequent in Campania; but that night the shocks were so violent that everything felt as if it were not only shaken but overturned.

By dawn the light was still dim and faint. The buildings around us were tottering, and the open space we were in was too small for us not to be in real danger if the house collapsed. That finally decided us to leave the town. We were followed by a panic-stricken mob of people who hurried us on our way by pressing hard behind. Once beyond the buildings we stopped, and there we had some extraordinary experiences which thoroughly alarmed us. The carriages we had ordered brought out began to run in different directions though the ground was quite level, and would not remain stationary even when wedged with stones. We also saw the sea sucked away and apparently forced back by the earthquake; at any rate it receded from the shore so that many sea creatures were left stranded on dry sand. On the landward side a fearsome black cloud was rent by forked and quivering bursts of flame, and parted to reveal great tongues of fire, like flashes of lightning magnified in size.

Soon afterward the cloud sank down to earth and covered the sea. It had already blotted out Capri and hidden the promontory of Misenum. Then my mother implored me to escape as best I could, . . . but I refused to save myself without her, and grasping her hand, forced her to quicken her pace. Ashes were already falling, not as yet very thickly. I looked around. A dense black cloud was coming up behind us, spreading over the earth like a flood. . . . A gleam of light returned, but we took that to be a warning of the approaching flames rather than daylight. However, the flames remained some distance off. Then darkness came once more and ashes began to fall again, this time heavily. We rose from time to time and shook them off, otherwise we would have been buried and crushed beneath their weight.

At last the darkness thinned and dispersed like smoke. Then there was genuine daylight and the sun was actually shining, but yellowish, as it is during an eclipse. We were terrified to see everything changed, buried deep in ashes like snowdrifts. We returned to Misenum where we managed as best we could, and then spent an anxious night . . . for the earthquakes went on.

Pliny the Elder is known for his great surviving work, the *Natural History*, a monumental encyclopedia of everything he could find that intrigued him. Written about the year 77 and divided into 37 "books," it abounds in recitations of obscure detail and petty history, and relates many tales that must have been suspect at the time, but it also contains accurate technical discussions of great scientific value. It has been acclaimed by some and derided by others, but it remains today the most important store of the scientific knowledge of the ancient world. Geological subjects are covered mostly in the last five books (XXXIII–XXXVII). Here Pliny provides much insight into the mineral industry of Roman times, with broad discussions of mining and smelting of metals, the value of earths for colors and ceramics, and the occurrence and uses of a long list of minerals and gems. He knows that mercury is poisonous and recommends alum as an antiperspirant. He gives an accurate account of the art of glassmaking, covers the ingredients in detail, and recognizes obsidian as a natural glass. A report on foundry practice leads him into digressions on sculpture and the art market, and the paragraphs on color grinding are followed by several essays on painting and art history. He covers the making of iron and the uses of steel in crime and in war and advocates arms control. He was a historian with a soul.

Pliny's *Natural History* was printed early and often. The first edition appeared in Venice in 1469 and was followed by a stream of other editions from dozens of presses. An excellent English translation by Philemon Holland (1552–1637) was published in London in 1601 and reprinted in 1634. Educated Europeans of the sixteenth, seventeenth, and eighteenth centuries read their Pliny—although it seems doubtful that many would have read him from cover to cover. Philemon Holland's translation runs to almost 1500 folio pages and weighs about 10 pounds.

The Romans had learned science eagerly from the Greeks, and they used it well in their empirical technology if Pliny is any example. They were well equipped and well organized when they marched north and west through Europe. Beginning with the conquest by Julius Caesar (?102–44 B.C.), the Legions maintained their *pax Romana* by force of arms, but they also carried communications and technology, their culture and their manuscripts with them, as civilian settlements followed the invading armies. Even a casual visit to Chester in England, Nîmes in France, or Trier in Germany, to mention three spots familiar to American visitors, gives a measure of the extent of Roman activity in northwestern Europe. It is easy to imagine the commerce, the wealth, and the intellectual endeavor commensurate with the great architectural landmarks.

CHAPTER TWO

The Middle Ages

MONASTERIES

When the network of the Roman Empire ultimately collapsed, the prosperity that was based on Roman organization soon decayed and Europe fell on hard times. The one great Roman force that survived the collapse of the Roman order was a way of life which the caesars never dreamed would be their most permanent export: Christianity. It spread and caught on like nothing else the Romans had ever adopted into their culture. Christianity had been spread by the legions, and it remained like a universal blanket when the legions melted away. Although post-Roman Europe was too poor and too preoccupied with the Christian hereafter to permit itself the luxury of science, numerous monasteries sprang up and became centers of stability and learning.

The monks copied, copied, and recopied the literature salvaged from the Romans, and thus prepared the way for European intellectual development. From humble beginnings in the fifth and sixth centuries, the monasteries grew to become the major cultural force in Europe for several hundred years.

ISLAMIC SCIENCE

While Graeco-Roman ideas were being copied and codified in the monasteries of Europe, they were also finding a fertile and in some ways more dynamic expression in the emerging world of Islam. Arabic culture flowered from about the eighth to about the thirteenth century in a vast empire that stretched from India to Spain. Science grew in the brutal splendor of the Arab courts, fostered as an ornament to the dynamic vanity of the warlords. Besides their great concentrations of wealth, a respect for erudition, and a remarkable degree of intellectual freedom, the Arabs also brought a new technical contribution to the growth of ideas: the invention of paper.

The art of papermaking had developed in China at about the time of Christ, spread west into Persia by the eighth century, and was carried by the Arabs into North Africa and Spain. From there it found its way to France and Italy in the thirteenth century, thus setting the scene, as it were, for the communications explosion that came with the development of printing.

The availability of paper made possible the growth of large private libraries at the Arab courts, but incessant warfare brought periodic destruction, with the result that few libraries survived for long. Mongol and Turkish invasions of the fourteenth and fifteenth centuries finished the destruction after the sacking of Baghdad in 1258. If only a small fragment of the Graeco-Roman literature was preserved by the monastic scribes of Europe, little more remains of the vast manuscript collections of the Arabs.

Islamic culture did more than transmit the ancient knowledge. The Arabs assimilated what they had learned from the Greeks and the Romans and then went beyond it with new mathematical methods, new experimental observations, and newly developed instruments. They translated the *Almagest* and for generations their mathematicians and astronomers continued to improve its methods and its data (without questioning the geocentric concept). The names of al-Battani (Latin *Albatenius*) in al-Raqqa on the upper Euphrates in present Syria (before 858–929), the Banu Musa (the brothers Muhammad, Ahmad, and al-Hassan, in Baghdad, ninth century), al-Zarqali (or Azarquiel) in Toledo and Cordoba (?–1100), and al-Bitruji (Latin *Alpetragius*) in Seville (late twelfth century) are prominent; their efforts led to the *Toledan Tables* and, somewhat later (about 1140), to the *Marseilles Tables*. These were the basic sources of astronomical data until Alfonso X of Castile (Alfonso el Sabio, 1221–1284) assembled his committee of savants in Toledo and they compiled, still on the basis of al-Zarqali's revision of the *Almagest*, the now famous *Alfonsine Tables*. These tables were over 200 years old when they first found a printer in Venice in 1483 in a Latin translation, but they were reprinted many times after that and were used widely in many ways, including some attempts to calculate the age of the Earth.

Beginning with al-Khwarizmi (before 800–after 847), Islamic astronomers added many new determinations of latitudes and a number of new longitudes to the improved tables appended to their astronomical compilations. The places they charted included not only cities of the Arab world but also mountains and points describing bodies of water. Ptolemy had constructed projections suitable for showing his world on flat maps, and he chose the Canary Islands, in the extreme west, as the place for the zero meridian. Reasonably accurate Ptolemaic maps of the Arab world with longitude measured east from the Canaries probably existed, but because there was no way of reproducing them reliably, they always were rare. Wood-engraved Ptolemaic maps reached wide distribution only after the printing press had taken over.

Arab astronomers did not concern themselves only with planetary problems, and Arab science generally had a very wide range. Al-Sufi (Latin *Azophi*, 903–986) is known for his *Book on the Constellations*, a catalog of the fixed stars (see Hyde, 1665), and Ibn al-Haytham (Latin *Alhazen*, 965–ca. 1040) for his *Optics*. Abu Ali al-Husayn Ibn Abdallah Ibn Sina (Latin *Avicenna*, 980–1037) was a far-ranging genius who held high positions and is remembered as a physician, philosopher, and encyclopedist. He also wrote some important geology, as we shall see.

The work of the Arabs had a tremendous influence on the development of European science because the last stronghold of Arabic learning was in Spain, where many Arabic manuscripts were being translated into Latin. Numerous scholars were involved, but Gerard of Cremona (ca. 1114–1187), an Italian working in Toledo, was by far the most productive translator and editor, and his name now dominates that literature.

Only a few of the Arab writings relate directly to geology, but they show advances from the geology of the Greeks—new measurements, new field observations, and better understanding. Ibn Sina (Holmyard and Mandeville, 1927) discusses the transitory nature of land and sea, cites the marine fossils found in some rocks as proof that some parts of present land were once submerged, and observes that running water can erode deep valleys. He follows Aristotelian views on earthquakes and tries to explain the origin of sedimentary rocks on the basis of the old Greek idea that water and earth can change into each other. He observes that some clay is formed by the disintegration of older rock but thinks of other clay as primary. At one point he writes about the piling up of younger layers of sediment on older ones and states or comes close to stating the concept now known as the law of superposition (Steno, 1916, 1969), but it is difficult to be sure what he really meant because the passage is unclear in the surviving Arabic texts.

Ibn Sina came to exercise considerable influence in the early development of geology when some excerpts from his encyclopedic *Book of the Remedy* (also known as *The Cure of Ignorance*) were translated into Latin and found their way to several publishers in the fifteenth and early sixteenth centuries. They appeared in various guises: as *Liber de Congelatione* under Ibn Sina's name, under the name of a translator, and as *Liber De Mineralibus* with Aristotle given as the author (perhaps because the publisher thought the book would sell better that way). Whatever the way, Ibn Sina's geological ideas received fairly wide distribution.

A different fate awaited the work of another brilliant Persian and Ibn

Sina's contemporary, Abu Rayhan Muhammad Ibn Ahmad al-Biruni (973–after 1050). His geological work was probably more extensive than Ibn Sina's, but most of his manuscripts are lost and those few that survived did not find translators for almost 1000 years. His astronomical work is cited in the Islamic literature and he undoubtedly contributed to contemporary geology, but the extent of his direct influence on later development must have been small.

His life was typical of the Arab scientist of the time: His intellect was recognized early; he studied a diversity of subjects in his youth; then he carefully survived the fluctuating political fortunes and personal whims of the monarchs who alternately supported and tyrannized him. He was primarily an astronomer, but his major surviving work, *India* (al-Biruni, 1910) is an important historical description of that country, its culture, and its institutions. It is based on his probably involuntary travels with the conquering armies of Sultan Mahmud of Ghazna (died 1030) and is sprinkled with a few important geological observations. Al-Biruni reports that particle size decreases in the sediment of the Ganges from the mountains to the sea and relates that to the strength of the current (al-Biruni, 1910, v. 1, p. 198). Of tides he writes: "The educated Hindus determine the daily phases of the tides by the rising and setting of the moon, the monthly phases by the increase and waning of the moon; but the physical cause of both phenomena is not understood by them" (al-Biruni, v. 2, p. 105). Like Ibn Sina he also takes a dim view of alchemy, perhaps surprisingly for his time: "One of the species of witchcraft is alchemy, though it is generally not called by this name" (al-Biruni, v. 1, p. 187).

Similar asides in other surviving works show that al-Biruni also observed fossil shells high above the sea and that he accepted them as remains of former sea life. He concluded that the evidence in the rocks indicates changes that could have developed only in very long periods of time. Later, in the reign of Mahmud's grandson Mawdud (1040–1048), al-Biruni wrote *Gems*, a treatise on the nomenclature, physical properties, and sources of precious and semi-precious stones and metals that includes precise determinations of density. That makes him the author of the third known text in mineralogy, but unlike his predecessors Theophrastus and Pliny, he did not have a chance to influence the mineralogical thinking of the Renaissance. *Gems* was first translated in 1963—into Russian (al-Biruni, 1963).

In recent decades, al-Biruni has become the object of veneration in his native Uzbekistan (Uzbek S.S.R.), but the depth of his scientific understanding should not be overestimated. He made many accurate astronomical observations and he studied fossils, but he did not anticipate either Newton or Darwin. Like the Hindus (but unlike Newton) he also did not really understand the origin of the tides. He remained well within the framework of his time, indebted to the Greeks for most of what he believed to be scientifically true.

THE BIRTH OF UNIVERSITIES

Out of the monastic culture grew a need for intellectual communication and professional training. Circles of learning sprang up, gradually developed into colleges, and combined to form universities, on a very modest scale at first, but they continued to grow. The Church of Rome may have had its occasional political troubles, but its centrality and the general acceptance of Latin as the universal language of learning fostered the diffusion of culture across national boundaries. By the mid thirteenth century important universities were active in Salerno, Bologna, Salamanca, Montpellier, Paris, Oxford, and Cambridge, dedicated primarily to theology, of course, but also concerned in varying degrees with medicine, law, and in a very small way with natural philosophy. Arab culture flowed into Europe through Spain and Aristotle was rediscovered. In Paris he was briefly condemned as pagan (from 1210 to 1235), but that did not last and his writings were widely studied.

Two English prelates, Robert Grosseteste (ca. 1168–1253) and Roger Bacon (ca. 1219–ca. 1292), plus a German nobleman, Albertus Magnus (ca. 1200–1280), and his Italian student, Thomas Aquinas (ca. 1225–1274), were gaining recognition by bringing Aristotle to the western world. They were theologians, above all, and their interest in the natural sciences was minor, but their great importance to geology lies in their indirectly preparing the ground for the growth of new ideas. Their views soon ossified into unalterable scholastic texts and their Aristotle effectively became a straw man to be attacked through new inquiries. The first geological ideas of the Renaissance arose from doubts of the Aristotelian corpus. It was the growing opposition to the rigidity of Aristotelian scholasticism that fueled early thinking about the nature of the Earth.

CHAPTER THREE

Geology Was Born in Italy

SHELLS IN THE HILLS

The distance from the mountains to the seashore is very small in Italy and the fossils in the hills are obviously similar to the shells on the beaches. How did the creatures of the sea become encased in the limestone? Leonardo da Vinci (1452–1519) was not the first Italian to consider the question—Giovanni Boccaccio (1313–1375) had written in 1370 that shells found in Tuscany were the remains of former sea life—but da Vinci was very likely the first to think it through and give the right answer. The marine origin of the fossils was perfectly clear to him. He wrote it down in his notebook and went on to other things.

In India, al-Biruni had been a virtual prisoner of his sultan but was free to say what he thought about the origin of fossil shells. In Italy, 500 years later, Leonardo was the recognized genius of his age, courted by popes and monarchs alike, but he was personally safe only so long as he adhered to official orthodoxy. The conclusions he had reached about fossil shells were heretical, but he had no intention of making them public. He could have suffered serious consequences had he insisted on publishing some of the ideas in his notebooks. The book of Genesis states plainly that God separated land and sea on the third day and created all fowl and water life on the fifth. Shells found on land thus obviously could not be remains of former sea life. The substance of Holy Writ was not open to amendment.

The Gutenberg Bible was being printed when Leonardo was born, and within his lifetime hundreds of printers carried their art from Germany to Italy and throughout Europe. The printing press was creating its own public and the tremendous power of the new medium soon became obvious to the authorities. Tight controls inevitably followed. The Church demanded that every manuscript for the press be reviewed by a designated censor who had to certify that there was no objection *(nihil obstat)* before an official permission to print *(imprimatur)* could be granted. Many statements in Leonardo's notebooks could not have passed such an examination in his time.

With empirical considerations of fossils so severely circumscribed, their study was bound to take some curious directions. Andreas Libavius (1560–1616), German alchemist and physician, defended the notion that fossils grow in rocks from some kind of seed. The idea may have come from Seneca and those fish-in-the-mud stories attributed to Aristotle and Theophrastus; and it was seriously discussed for almost 200 years more. The concept of a *plastic virtue* and *lapidifying juices* arose from Albertus Magnus' interpretation of Avicenna (in *De Mineralibus et Rebus Metallicis*, book 1, chapter 9), repeated (without credit) by Girolamo Savonarola (1452–1498) in *Compendium Totius Philosophiae* (Savonarola, 1534).

Albertus' earnest effort to understand the process of fossilization without having to assert the organic origin of fossils was gradually corrupted into theories explaining fossils as accidents or *sports of nature (lusus naturae)*, some humorous but others malevolent, conceived by Satan to confound the minds of mortals. A safe attitude was adopted by Michele Mercati (1541–1593) in his catalog of the Vatican collection of minerals and fossils, which was completed in 1574 but published only much later (Mercati, 1719). He proposed that fossil remains were objects *sui generis* (of their own kind) not related to anything in particular and presumably created by God together with everything else. That concept survived well into the eighteenth century.

Very strange properties were ascribed to two classes of "fossils"—*cerauniae* (thunderstones) and *glossopetrae* (tonguestones). They were supposed to fall from the sky in thunderstorms and other calamities, and their powers were usually reported as evil. The origin of these myths is shrouded in antiquity, but in old illustrations (e.g., Gesner, 1565) it is easy to recognize that *cerauniae* were belemnoids, some smooth echinoids, and polished neolithic implements. In northern Germany, silicified belemnoids from the Chalk are still called *Donnerkeile* by the countryfolk, and the corresponding term *thunderbolts* still crops up in England. *Glossopetrae* were common on Malta and became vaguely associated with St. Paul's shipwreck there. It was said that they were the fossilized tongues of snakes and dragons, and that view persisted long after Konrad Gesner (1516–1565) had illustrated them (Gesner, 1565), suggesting that they were fossil sharks' teeth, based on their similarity to the teeth of a modern shark which had been sent to him by a pharmacist from Antwerp.

COLLECTIONS AND MUSEUMS

The Renaissance brought new interest in nature and a new twist to the ancient custom of building collections. Weapons, treasures, and art objects had always been collected, but the new directions were *naturalia* and prehistory, and that included minerals, fossils, shells, feathers, dried plants, mounted animals, ancient pottery and other antiquities, including tools, weapons, and coins. Private collections, the "cabinets" of the virtuosi, began to grow into extensive accumulations in the sixteenth century. Many were dispersed again, but some of the largest became institutionalized under various governmental sponsorships and ultimately became museums.

Some of the virtuosi wrote detailed catalogs of their own collections and others hired curators to do it for them. Many such catalogs were published and some are richly illustrated, so that now we have a very good picture of what these "cabinets" contained. There was, of course, an element of vanity in much of this collecting (see, e.g., Beringer, 1963), but many of the collectors were actually scientists, and their collections served as microcosms for the scientific study of nature. The cabinets were usually available to invited visitors, but their printed catalogs were much more widely distributed and thus provided indirect access to many scientifically inclined readers. These catalogs were the backbone of the geological literature of the Renaissance.

Among the earliest of those catalogs are Johann Kentman's (1518–1574) of Dresden (Kentmannus, 1565), Konrad Gesner's of Zurich (Gesner, 1565), the three catalogs of Francesco Calzolari's collection in Verona (Olivi, 1584; Ceruti and Chiocco, 1622; and Moscardo, 1656), the catalog of the papal collection (Mercati, 1719), and the enormous accumulation of Ulisse Aldrovandi (1527–1605) in Bologna, which was cataloged in his own hand in 187 volumes. The published version runs to 13 large folio volumes, one of which deals with geological material (Aldrovandi, 1648). It is a vast encyclopedia of what was then known about animal, vegetable, and mineral nature.

THE DILUVIAL SCHOOL

Benedetto Ceruti and Andrea Chiocco, in their catalog of the Calzolari collection (Ceruti and Chiocco, 1622), quote the opinion of Girolamo Fracastoro (1483–1553), philosopher and physician, that fossils are remains of real animals. Fracastoro's ideas may have been influenced in discussions with Leonardo, and they were aired in print only long after both he and the Master were safely deceased. Bernard Palissy (1510–1581), the French potter and thinker, wrote that fossil shells had lived and were petrified on the same spot where they are now found (Palissy, 1580). He also observed that many fossil forms are quite different from modern species. In another catalog of the same Calzolari collection, Olivi (1584) championed the *lusus naturae* hypothesis but admitted that some think that *glossopetrae* are sharks' teeth. Fabio Colonna (1567–1650), a Neapolitan physician, openly and clearly said so (Columna, 1616).

Girolamo Cardano (1501–1576), the neurotic inventor (of the universal joint) and a great mathematician, asserted an idea that was novel for his time: All running water comes from rain, and rain is caused by the evaporation of the sea (Cardano, 1550). He quotes Pausanias and adopts Avicenna's view that mountains are made by erosion and that marine shells in the hills are proof that high land was once submerged. These ideas all recall those of Leonardo, but how Cardano could have had access to them is debatable. His father had known Leonardo, but young Cardano was only 18 when the Master died, and the notebooks remained hidden as far as one knows. (For a contrary view see Duhen, 1906.) The discussions with Fracastoro did not come to light until 1622 (Ceruti and Chiocco, 1622).

It seems likely that Palissy had cribbed from Cardano, but worse than that, he tagged him with the dubious honor of having founded the Diluvial School, which sought to explain all fossils in terms of the Noachian Deluge. Cardano had not invented the diluvial doctrine, nor had he preached it. The concept goes back at least to the early Greeks, to Xenophanes (ca. 570–ca. 480 B.C.), who thought that great floods destroy all life, and to Anaximander (ca. 611–ca. 547 B.C.), who believed that all life originated in a flood (Kirk and Raven, 1962). Many writers have cited the remark by Pausanias noting fossil shells high above the sea in the limestone around Megara (about 20 miles west of Athens), but actually that observation is barely a footnote. "It is white, softer than any other, and has sea shells all the way through it. That is what the stone is like," wrote Pausanias, and that is all he wrote about it (Pausanias, 1971, v. 1, p. 124).

The diluvialist concept grew from these ancient roots through the Renaissance, flourished in the seventeenth and eighteenth centuries, and produced a tremendous literature. In America it was scientifically defended almost to the end of the nineteenth century, and it is still being trumpeted pseudoscientifically by some religious groups.

THE METALLICK TRADITION

By the beginning of the sixteenth century, the mining business in central Europe was solidly in the hands of the German mining fraternity, even in areas that were not ethnically German. Exploration was highly successful and production was booming. Mining communities sprang up in Iglau (now Jihlava), Joachimsthal (now Jáchymov), and Kuttenberg (now Kutná Hora) in Bohemia; in Freiberg and Chemnitz (now Karl Marx Stadt) in Germany; in Schemnitz (now Bánska Štiavnica) in Slovakia; and in Temeśvar (now Timişoara) in Romania. Among other metals, the mines produced silver for the coinage of the various sovereigns, and in return for this valuable service the miners were able to extract from the central governments far-reaching concessions that gave local administrations a large measure of political freedom and fiscal independence.

The prosperity of the mining towns and their politically privileged status combined to make them attractive. The miners needed physicians, teachers, and preachers, and the governments of their towns had

FIGURE 3.1. Figure of a bearded miner with pick and ore hod, holding a candlestick in his teeth. Brass in Newland Church, Gloucestershire; the plate measures 30 by 18.6 cm and its provenance is unknown, but the style implies a date in the fourteenth century. That would make it the oldest known image of a western miner.

no trouble filling these posts with high-caliber people. A new intellectual element found itself at home in the midst of the mining communities and a genuine mining literature began to develop. Ulrich Rülein von Calw, or Kalbe (1465/1469–1523), renowned engineer and long-time physician in Freiberg, wrote the first mining manual, the *Bergbüchlein* (Rülein, 1500). It was a modest effort, containing little more than digests of ancient traditions of the origin and distribution of metallic ores (based largely on Pliny), and various medieval notions of astrology and alchemy as related to mining. Those stories could not have been of much practical use to the mining fraternity, but Rülein also included elementary instructions for opening a mine, which may have been the main reason for the *Bergbüchlein's* popularity. It saw many editions, some of the later ones accompanied by *Probirbüchlein* (assay manual) not written by Rülein, and by summaries of mining law, as it was then developing.

The first important Renaissance work on metallurgy was written by Vanoccio Biringuccio (1480–ca. 1539), a Sienese who had studied in Germany and then served in various engineering and metallurgical positions in Siena, Venice, Florence, and Rome. He is remembered for his only book, the *Pirotechnia*, published posthumously (Biringuccio, 1540, 1943), and reprinted several times. It is an important and original treatise on minerals, assaying, parting, smelting, foundry practice, and other metalworking, based on a purely empirical approach and extensive practical experience. It had a far-reaching influence on European mineral technology.

AGRICOLA

The mining community of Joachimsthal was just 11 years old when Georg Bauer (1494–1555) was hired as town physician in 1527. He was highly recommended as a doctor, but apart from that he was exceptionally well read in the classics, highly inquisitive in technical matters, and young enough to be open to new ideas. In his spare time he wrote books in Latin, and signed them Georgius Agricola. He recognized his marvelous opportunity to study the mining business at its source and took full advantage of it. He gained the confidence of the miners and learned their special vocabulary. In 1534 he moved to another mining town, Chemnitz, practiced medicine there, made some fortunate investments in mining and smelting, and served in various official and diplomatic capacities, including four terms as burgomaster.

The books he wrote constitute the first comprehensive technical record of the mining industry. He studied all phases of the business in the field—from preliminary prospecting through every detail of mine oper-

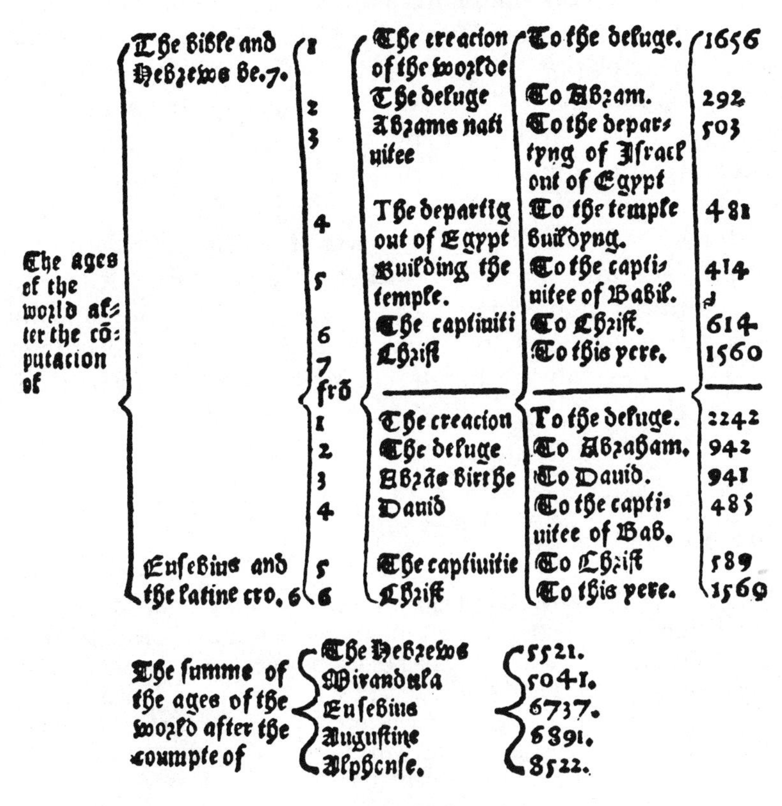

The computācion of the agēs of the worlde.

The ages of the world after the cōputacion of

The Bible and Hebrews be. 7.

1	The creacion of the worlde	To the deluge.	1656
2	The deluge	To Abraham.	292
3	Abrahams nativitee	To the departyng of Israel out of Egypt	503
4	The departig out of Egypt	To the temple buildyng.	481
5	Buildyng the temple.	To the captivitee of Babil.	414
6	The captivitie	To Christ.	614
7	Christ	To this yere.	1560

frō

Eusebius and the latine cro. 6

1	The creacion	To the deluge.	2242
2	The deluge	To Abraham.	942
3	Abrahās birthe	To David.	941
4	David	To the captivitee of Bab.	485
5	The captivitie	To Christ	589
6	Christ	To this yere.	1560

The summe of the ages of the worlde after the coumpte of

The Hebrewes	5521.
Mirandula	5041.
Eusebius	6737.
Augustine	6891.
Alphonse.	8522.

FINIS.

FIGURE 3.2. Calculations of the "age of the world" were made long before Bishop Ussher. (Cooper's Chronicle, 1560)

ation to the refining of the ultimate product—and critically reviewed all the information he could find in the literature. It would be an exaggeration to say that he created mineralogy, mining engineering, and economic geology, but he did much more than report on the state of the art of mining.

While traveling when he was still a medical student, Agricola had met the great humanist Erasmus (?1466–1536), editor of the famous Froben Press in Basel, and apparently impressed him with his ability. Some years later Erasmus wrote a letter of introduction for Agricola's first geological work, *Bermannus*, and that assured its publication by Froben (Agricola, 1530). The book introduced Agricola as a mining scholar and established his lifetime association with the Basel publisher.

Bermannus is a brief introduction to mineralogy and the mining business set in the form of a dialogue between an experienced miner (modeled after one of Agricola's best informants in Joachimsthal, Lorenz Berman) and two philosophers wishing to learn about the business. Agricola himself had only about two years experience when he wrote it, but he continued to work on the mineralogical parts of it and much later wrote *De Natura Fossilium* (Agricola, 1546, 1558), which greatly expands and improves the treatment of the subject. It is a mildly critical and thorough summary of all ancient literature on rocks, minerals, fossils, and pseudofossils, relying heavily on Pliny and paying great homage to Theophrastus. Even marginal details such as Herodotus' story of the great bronze vase of Croesus are included in the brilliant display of Agricola's classical erudition, but the book is ultimately founded in his direct empirical studies and therein lies its great value. It is a first attempt at a comprehensive system of mineral classification and the first step forward from the wild redundance and synonymy of ancient nomenclatures of rocks and minerals. It was a point of departure for the development of mineralogy.

The Bronze Vase of Croesus

In early 1953, a bronze vase over 5 feet tall, including base, and about 4 feet in diameter, was excavated from a grave in the upper valley of the Seine River near Vix, France. The body of the jar, about a sixteenth of an inch thick, is hammered from a single piece of bronze. Magnificently ornamented with tracery, a bronze frieze of figures, ornate handles and lid, it is obviously Greek workmanship. The grave, that of a Celtic princess, was dated by the archaeologist in charge, Rene Joffroy, at about 500 B.C.

Why was this magnificent Greek artifact found in the Cote d'Or region of France? Herodotus, in the century after the princess' burial, describes (1-70, Rawlinson's translation) the efforts of Croesus, the king of Lydia in western Asia Minor, to form a defense alliance with the Lacedaemonians. He sent gifts to Sparta and they, in turn " . . . had a huge vase made in bronze, covered with figures all around the outside of the rim, and large enough to hold three hundred amphorae, which they sent to Croesus . . . " If this be that vase, it never reached its destination. It may have ended up as a bribe to a Celtic princess who controlled the overland route to Greece for tin mined in the lower Loire valley of France and, obliquely across the Channel from the mouth of the Seine, the mines and placers of Cornwall. (The Greeks had no tin deposits and had to import huge quantities for their bronze making.)

The vase is now exhibited in the small museum in Châtillon, France.

Agricola's geological views are covered in *De Ortu et Causis Subterraneorum* (Agricola 1546, 1558). It is highly critical of the supernatural but presents no great departure from classical ideas, such as the four elements and the interaction of hot and cold. It gives much credit to Ibn Sina, perhaps mainly to support Agricola's own observations of the power of erosion by frost and running water, and presents an original attempt at classification of ore deposits.

The book for which Agricola is best known is *De re Metallica*, completed just before he died and published posthumously (Agricola, 1556, 1912). It is a rambling and all-inclusive summary of the state of the art of mineral exploration, mining, drainage, ventilation, haulage, ore dressing, smelting, refining, and glassmaking—made particularly valuable by the many lively woodcut illustrations by Hans Rudolf Manuel Deutsch. It is not wholly original (the sections on metallurgy and glassmaking rely heavily on Biringuccio), but it presents the subject in a magnificently comprehensive way. It was frequently reprinted and remained the basic reference of the mining industry for two centuries.

Like the erudite Renaissance man he was, Agricola wrote in Latin even though that language was particularly ill-suited to his purpose. He could use Pliny's terms, as far as they went, but his empirical observations were based on German sources, and the complicated and precise technical terminology of the German mineral industry had no Latin equivalent. Agricola was forced to invent a whole new Latin vocabulary for many of the common terms and elaborate transcriptions for the rarer ones. As a result, some of his text was difficult to understand even by the contemporary reader. To make things worse, *De re Metallica* was translated into German (Agricola, 1557) by Philip Bechius, a Basel physician unfamiliar with the specialized German of the miners. In that form it is even more obscure than the Latin original. He had tried to compile a Latin-German glossary (Agricola, 1546, 1558) but it was too brief and Bechius did not use it.

FIGURE 3.3. "I have known . . . more than fifty persons who used this simple instrument to find waters, minerals, & hidden treasures, and it truly turned in their hands." (Vallemont, 1696, p. 9)

The new intellectual interest in mining produced some remarkable religious books, of which perhaps the most interesting is *Bergpostilla oder Sarepta* by Agricola's close friend Johann Mathesius (1504–1565). Mathesius was the parish priest in Joachimsthal and the *Bergpostilla* (Mathesius, 1562) is a collection of 20 sermons analyzing all the biblical references to minerals and mining, and interpreting them in the light of contemporary mining experience. It must have been very popular because it was reprinted many times, and the various editions contain up-to-date historical summaries and tables of annual production figures for each of the Joachimsthal veins—a fine testimonial to theological pragmatism in the mining community, and a point for the booksellers.

Testing and assaying of ores were difficult technical problems for the miners of Agricola's day. A bewildering array of empirical techniques had been developed since antiquity (shrouded in many layers of convention and superstition), whereas the lack of chemical knowledge and the absence of a standard nomenclature impeded rational development of the art. The first serious attempt to bring order to this maze of redundant and conflicting traditions was made by Lazarus Ercker (1530–1594), the superintendent of mines for Rudolf II of Austria. He wrote a comprehensive assay manual (Ercker, 1574) describing not only the ores and the techniques for assaying them, but also the preparation of reagents and the construction of analytical apparatus.

The modern reader may have trouble trying to follow Ercker's procedures, but in their day they represented the sum of the art of assaying and his manual dominated the field for 150 years (Ercker, 1951). It appeared in eight editions in German and was translated into English by Sir John Pettus (1613–1690), the Deputy Governor of the Mines Royal for more than 35 years. He did it to pass the time while serving a term for debts in Fleet prison.

CHAPTER FOUR

A Solid Within a Solid

THE NEW THEORY

Scientific discussions in the late sixteenth and early seventeenth centuries were dominated by the uproar over the heliocentric theory of Nicholas Copernicus (1473–1543). He was a canon of the cathedral and spare-time astronomer in Frauenburg in East Prussia (now Frombork in northern Poland), and the theory was his life's work. He had allowed his idea to circulate in preliminary form as an unsigned manuscript for some 30 years, and the indirect reports he had received from mathematicians and astronomers had been favorable, but no one had said much in public. During the last year of his life he was persuaded to allow it to be published in far-away Nuremberg, but by that time he was too ill to supervise the venture himself, and the event brought him little pleasure.

The task of seeing *De Revolutionibus* through the press (Copernicus, 1543) fell to an overzealous Lutheran cleric, Andreas Osiander (1498–1552), and he slipped in a preface he had written (but not signed) to the effect that the whole thing was just a hypothesis and "may not be true." It is not clear whether Osiander was trying to make the work more palatable to theologians or whether he was just injecting his own views. Copernicus was annoyed when he saw the spurious preface, but it may have helped to keep the clergy at bay, at least for some years.

Scientifically, the Copernican concept was a bold step forward, but the theologians could not see it that way, and for once the Catholic and Lutheran clergies found something to agree on. Joined by old-line scientific conservatives, they mounted a violent opposition to the heliocentric theory that ultimately grew entirely out of proportion with the theological problems involved. The Catholic church placed the *De Revolutionibus* on the Index of Prohibited Books in 1616 and kept it there until 1835. The Copernican "heresy" was denounced with a vehemence never before accorded to a scientific idea.

The drumbeat of theological objections to the theory, in the long run, served mainly to publicize it and to make people take positions, pro or con. Tycho Brahe (1546–1601) had doubts about it, but the superbly accurate observations he made of planetary orbits helped ultimately to confirm it. Johannes Kepler (1571–1630), Brahe's young friend and intellectual heir, became an enthusiastic Copernican and through his research was able to remove almost all astronomical objections that had been raised to the heliocentric theory in its original form. Once it was understood that the orbits of the planets around the sun were elliptical rather than perfectly circular, everything fell into place, but that was a scientific explanation and it did not alter the theological position of the Church. Galileo Galilei (1564–1642) became the most prominent defender of Copernicus (Galilei, 1632) and as a result was brought to trial and condemned to life imprisonment. The sentence was commuted and he never served time in jail, but his freedom was restricted for the last decade of his life. Had it not been for powerful friends, he could have fared much worse.

The publicity of Galileo's trial led many scientific thinkers to increase their guard and to publish with greater caution, but it also made clear that the Copernican system was valid and that the objections to it were other than scientific. Thus it stimulated the more or less clandestine discussion of scientific concepts that could be at variance with theological opinion. Prominent among them was the debate over the origin of fossil remains.

Gesner (1565), Palissy (1580), Imperato (1599), Columna (1616), and Ceruti and Chiocco (1622) had all said in print that the fossils in the hills were remains of former sea life, but that was not the official view. They were the voices of a small minority and any one of them could have found himself in serious trouble had he chosen to push the idea too strongly. It was prudent for the naturalists to be cautious and keep their studies in a low key, but the question of the organic origin of fossils was alive and being examined.

Such was the mood in Italy when a young Dane arrived in Florence at the invitation of Grand Duke Ferdinand II of Medici. His name was Niels Stensen (1638–1686), latinized to Nicolaus Stenonis or Steno. He had studied in Copenhagen and in Leiden, discovered the excretory duct of the parotid gland (still known as Stensen's duct) in Amsterdam, defended a thesis there on hot springs, and ultimately received a doctorate in medicine from Leiden (in absentia) while dazzling his elders in Paris with his command of anatomy and his virtuosity in dissection. He also spent time in Montpellier, and from there traveled to Florence. Whatever else may have been said about his new sponsor, Ferdinand II, few could deny that he had a nose for talent.

The new doctor immediately plunged into a multitude of anatomical

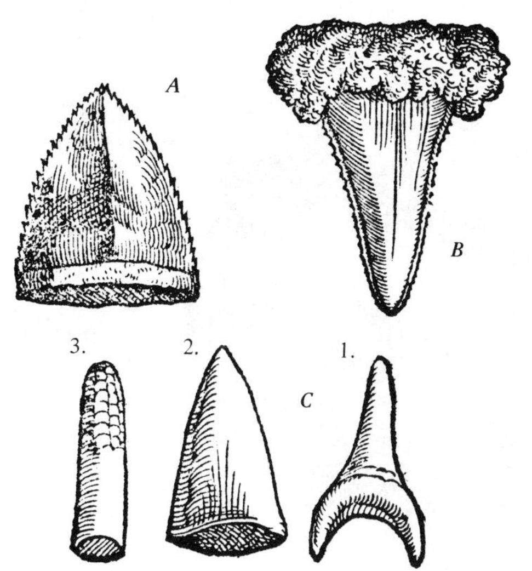

FIGURE 4.1. "A. B. C. are stones we call Glossopetrae, or the teeth of sharks . . ." (Gesner, 1565)

studies. In the fall of 1666, an unusually large shark was brought ashore near Livorno and the Grand Duke ordered the monstrous head of the beast brought to Steno for dissection. By that time Steno had completed a dissertation on the geometry of muscles. (It turns out to have been a scientific blind alley with a misapplication of mathematics to a biological problem—and it had already gone through the hands of the first censor.) The origin of *glossopetrae* or tonguestones, flattish roughly triangular objects often found embedded in rock, was still debated and the teeth in that shark's head seemed familiar to him. He went to work on the specimen.

Within two months his report was ready and he attached it to the treatise on muscles, to add variety, he said (Steno, 1667). It was a carefully worded piece of writing, apologetic, almost timid in spots, but leaving no doubt of the author's understanding not only of anatomy but also of the processes of sedimentation and fossilization. It made clear his firm conviction that *glossopetrae* are fossil sharks' teeth, and that the fossil shells found in rocks in many places are the remains of animals that once lived there. (For a thorough description of Steno's reasoning see Albritton, 1980, pp. 23–29.)

Then for more than a year Steno studied geology. He traveled through Tuscany looking at strata and collecting fossils. He visited private collectors and examined their treasures. He went to quarries and climbed into caves, and traveled to Elba to see the famous mines that still produce spectacular crystals of pyrite and hematite. In the midst of that activity, in November 1667, Steno's work was interrupted by two events: He was converted to Catholicism, and a month later he received a letter from his king, Frederick III of Denmark, offering a substantial

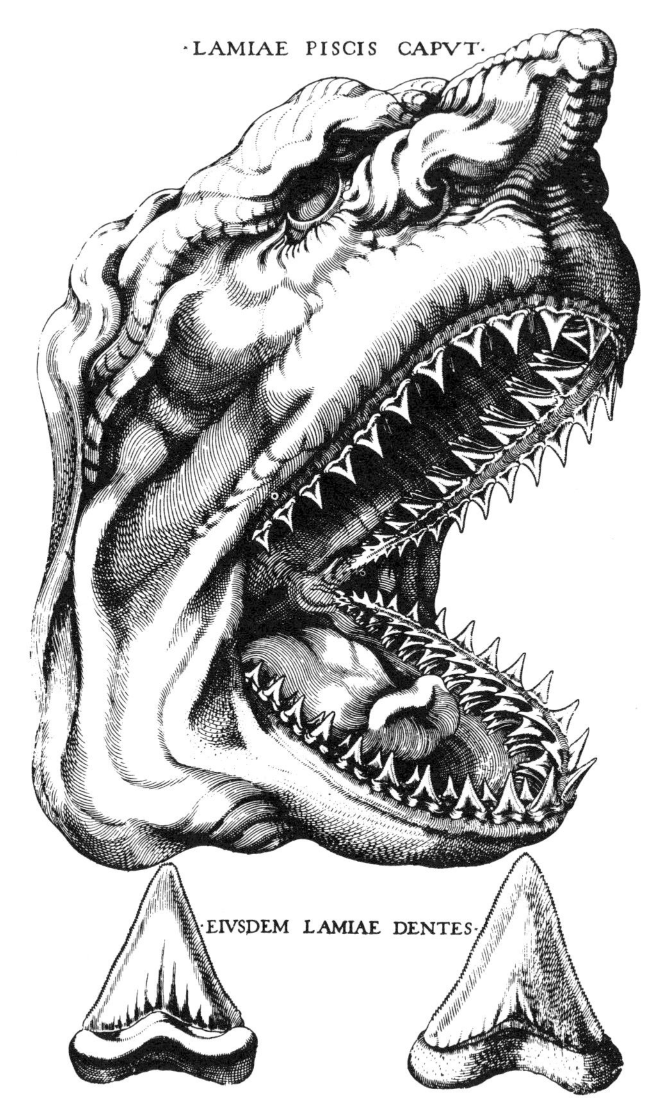

FIGURE 4.2. The head of a shark and its teeth. The drawing was borrowed from a then unpublished manuscript (Mercati, 1717) by Steno for his *Myologia*. (Steno, 1667)

stipend and ordering him to come home. Realizing that his research was about to be seriously disrupted, Steno composed a systematic summary of his results. He called it *De Solido Intra Solidum Naturaliter Contento Dissertationis Prodromus (Forerunner of a Dissertation of a Solid Naturally Contained Within a Solid)*, promising that this was just a brief harbinger of a larger work to come, but leaving enough hints to show that he knew that would never be.

He reduced geological theory to one fundamental question: How does

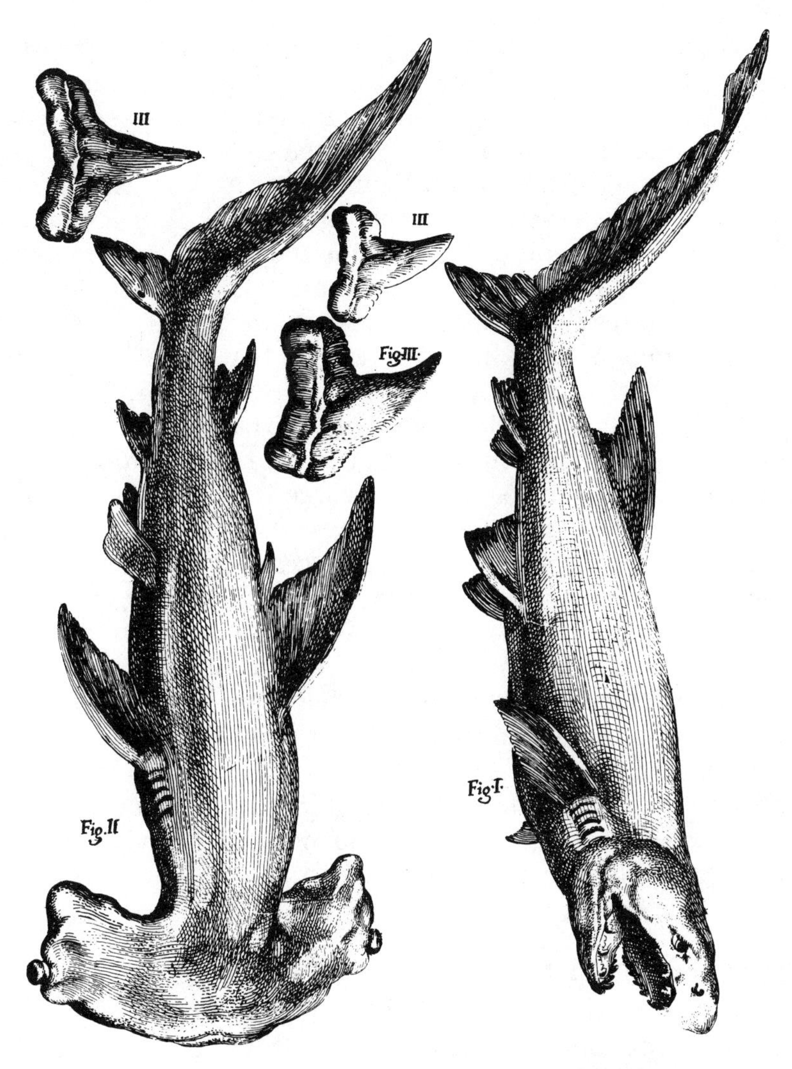

FIGURE 4.3. "Fig. I. Drawing of the entire Dog Fish . . . Fig. II. Hammer Fish, drawn from life, which is armed with teeth similar to many found petrified on Malta. Fig. III. Teeth of the Hammer Fish, which in variety, number of teeth, and any other characteristic of the mouth does not differ from the Dog, or similar fish." (Scilla, 1670)

a solid, like a shell or a crystal, become enclosed by another solid? The enclosed body had to be solid before the enclosing rock solidified. Extending the concept to rock strata, he outlined the process of sedimentation, analytically demonstrated the organic origin of fossils, and formulated the law of superposition. The stratum below had to be complete by the time the layer above it was being deposited, and strata now above that layer then could not yet have existed. He understood that metalliferous veins were younger than the rock that encloses them and that crystals could grow in fluid-filled cavities in solid rock. He described some characteristic forms of crystals, outlined their growth, noted fluid inclusions in them, and recognized that the angles between crystal faces are constant regardless of the size or shape of the crystal.

In the *Prodromus* he shows no need for any apology. He has made his observations and the picture is clear—even if his soul is not at peace. Firmly planted in his mind now was the seed of the inevitable conflict between pious interpretation of Genesis and the conclusions flowing so clearly from the results of his own perception.

The *Prodromus* was published in April of that year (Steno, 1669), but not without difficulty. The first censor was Vincenzo Viviani (1622–1703), the mathematician who liked to call himself "Galileo's last student," and was a vigorous defender of the Master's memory. He saw the importance of Steno's work and urged its publication. It seems likely that his appointment as censor was engineered for just that reason. The second censor was Father Francesco Redi (1626–1698), who had then just published his now famous experiments disproving the spontaneous generation of insects. He was a conservative churchman and he kept the manuscript for four months before he finally approved its publication. One can only guess what went on behind the scenes during that time. Steno himself had had second thoughts long before then, and in the end his devotion won out. He left the publication to Viviani, and the *Prodromus* was printed from a manuscript in Viviani's hand (Scherz, in Steno, 1969). Steno abandoned geology and spent the rest of his peripatetic life serving the Church.

GRESHAM COLLEGE

The discovery of America was a great boost for science; it produced a new frontier with new natural phenomena to be observed and described. It also demonstrated the usefulness of astronomy and mathematics in navigation and thus encouraged new research in both. The ancient little book *De Sphaera (The Sphere)* by John of Holywood (or Sacrobosco, died about 1256), which had been the basic text of astronomy and celestial navigation for three centuries, finally became obsolete and was replaced by a succession of more sophisticated texts and tables that were more useful to the mariner and also more accurate.

The compass needle was perfected and came into general use. Improved methods of navigation led to better charts, and better charts encouraged more navigation. Map making developed at a spectacular pace in the sixteenth century, particularly in the Netherlands. The blossoming British maritime trade produced new expansion in all directions and coincidentally new sources of wealth for the endowment of scientific endeavors.

The new philosophy of benefiting humanity through the teaching of science led Sir Thomas Gresham (?1519–1579) to endow a new college for bringing scientific education to craftsmen and merchants in London. Sir Thomas' name is now popularly connected with Gresham's Law—the principle that bad money drives out good—but in his time he was known simply as the richest commoner in England. It took some time and litigation to straighten out the provisions of his will, but Gresham College finally opened in 1597, providing classrooms and lodgings for a faculty of seven professors, one for each day of the week. (Theology was taught on Sundays, of course.) The will specified that the professors be unmarried, and should any one of them take a wife, his appointment would cease. That may have been a perfectly conventional stipulation at the time, but had Sir Thomas understood human nature as well as he did finance, he could have anticipated that this requirement would lead to more sin than virtue at Gresham College.

Whatever the attraction, Gresham College became the focal point of the London scientific scene. There was no great rush of tradesmen to the lecture halls, but gentlemen of scientific bent began dropping in with increasing frequency. John Napier (1550–1617), among other accomplishments, had perfected the concept of the logarithm (Napier, 1614); and Henry Briggs (1561–1630), professor of geometry at Gresham College (and later the first professor of mathematics at Oxford) published the first large decimal logarithmic tables in 1624. For about 350 years those tables and their descendants were the main computational aid for surveyors and navigators until displaced by small electronic calculators.

The new interest in navigation may have led William Gilbert (?1540–1603) to his experiments with the lodestone. Combining brilliant experimental technique with a healthy contempt for tradition and hearsay, he examined the basic principles of magnetism. He had a sphere *(terella)* made of a homogeneous piece of lodestone (naturally magnetic magnetite) and charted the shape of the magnetic field around it with small magnetic needles, demonstrating that it was similar to the observed field of the Earth (Gilbert, 1600). The Earth was like a giant lodestone, and Gilbert's concept of an all-pervading magnetic force field was a great step toward the later development of the gravitational field theory by Newton.

The inclination of magnetic needles from the horizontal had been noticed and accurately measured by Robert Norman in 1576. Little is known about him except that he sold nautical instruments in London and wrote a successful book on magnets (Norman, 1581) which went through many editions and was favorably cited by Gilbert. Measuring the deviation of the needle from true north (the declination) was more difficult because it involved repeated and accurate observations of the sun with respect to the needle to determine its angle with the astronomical meridian. William Borough (1536–1599), a retired naval commander and explorer, made those measurements in 1580 in his garden in Limehouse (now in East London), and they were published together with the inclination data in Norman's book. That was the beginning of magnetic observation in London.

Edmund Gunter (1581–1626), professor of astronomy in Gresham College, repeated Borough's measurements at Limehouse in 1622 and found a declination lower by 5 degrees and 25 minutes (Gellibrand, 1635). He could not explain the apparent discrepancy and assumed that Borough had been wrong. This may seem unkind, but was understandable at a time when statistical analysis of experimental data was still virtually unknown. [The experimental uncertainty (mean deviation) of Borough's data is actually only 4 minutes of arc and Gunter's 10 minutes.] Henry Gellibrand (1597–1636), Gunter's successor at Gresham, repeated the experiment in 1634 and found that the declination had decreased by almost 2 degrees more since 1622 (Gellibrand, 1635). (The mean deviation of Gellibrand's data is also only 4 minutes.) The variation of the Earth's magnetic field in time was real.

Such experiments, and many others in widely diverse fields, often involved many people and complicated loans of elaborate, privately owned equipment. They were conducted by gentlemen for the fun of it as much as for the science, and to provide material for discussions. This kind of high-level entertainment was eminently suitable to the tastes of a circle of largely amateur "philosophers" that gradually formed at Gresham College. On the evening of November 28, 1660, after a lecture by Christopher Wren (1632–1723), professor of astronomy since 1657, the 12 gentlemen present decided to form an organization for the advancement of science and to meet every week. A year and a half later the group was chartered as the *Royal Society*.

FOSSILISTS OF THE ROYAL SOCIETY

The Royal Society could have become just another amateur science club, except for the presence among the early members of six dedicated professionals. Robert Boyle (1627–1691), John Wilkins (1614–1672), and his best student, Wren, were present at the first meeting. Boyle soon brought in his inventive assistant Robert Hooke (1635–1703). John Ray (1627–1705) joined in 1667 and Isaac Newton (1642–1727) in 1672. It was a remarkable galaxy of talent and all of them took up geological

his apartment when he died were finished, and the accompanying drawings were ready for the engraver. The principal paper was dated 1668, the rest spanned the period from 1684 to 1699, and he could have published them all without much additional effort. Instead, he put them in a trunk and they were printed only in 1705 after the new Secretary of the Society, Richard Waller, had collected them into *A Discourse on Earthquakes,* which forms the major portion of the *Posthumous Works* (Hooke, 1705).

A few of Hooke's later geological papers are polemical and imply that he was attacked for his uncompromising stand on the organic origin of fossils. The opposition appears to have come from within the Royal Society, and that was the forum in which he had responded, but his replies were not particularly forceful. He seems to have felt no urge to convert the theologians or to fight them, and his unwillingness to publish may have come from a deep reluctance to expose himself in a public debate of this dangerously controversial issue.

The interaction of Hooke with Steno is obscure. Hooks's most important paper on fossils bears the date of September 15, 1668, and by that time Steno's *Prodromus* was already through the hands of the first censor. It seems likely that Oldenburg had reported the gist of Hooke's ideas to the Accademia del Cimento in Florence, together with much other news (Oldenberg, 1965–1972), and Steno may have seen the *Micrographia* before he finished the *Prodromus.* Hooke hardly could have failed to see the *Prodromus,* especially after it was published by the Royal Society, "English'd by H(enry). O(ldenburg)." Steno makes no reference to any previous author in the *Prodromus,* and nowhere in Hooke's writings is there a direct mention of Steno.

Hooke's experimental genius was unequaled in his century, but his administrative talents were slight. When Oldenburg died, he took over as the Society's secretary and it was not long before publications began to lag and correspondence was adrift. When Newton submitted his first paper on colors to the Royal Society in 1672, Hooke made some haughty comments about it that also happened to be wrong, and a life-long feud developed between those two towering egos. In their simmering hostilities both of them behaved badly, but Newton was the implacable one. He refused to be president of the Society as long as Hooke was alive and apparently delayed the publication of his *Opticks* (Newton, 1704) for the same reason. His influence is felt in the unflattering biography of Hooke written by Waller as a preface for the *Posthumous Works* (Hooke, 1705). An imposing portrait of Hooke still hung in the Society's meeting room at Gresham College in 1710, but vanished after the Society moved to new quarters later that year. The move had been pushed through by Newton, and the suspicion has been raised ('Espinasse, 1956, p. 13) that he may have had something to do with the portrait's disappearance.

Steno's *Prodromus* must have made an impression on the management of the Royal Society or Henry Oldenburg would not have taken the trouble to translate it, but it was not a big seller. It was remaindered after two years with the unsold sheets of Boyle's *Gems* and several other essays, with a new title page (Boyle, 1673). Still, coming from the Royal Society, Steno's ideas must have had fairly wide circulation, even if their reception may have been unenthusiastic. Few writers could bring themselves to quote Steno in print, but even so he fared better than Hooke, whose geological views were almost universally ignored until they were rediscovered with much fanfare (Raspe, 1763) almost a century after they had first been read before the Royal Society.

REMAINS OF FORMER LIFE?

Hooke had not succeeded in bringing his fossilist colleagues to recognize that the objects they called "fossils" included two altogether unrelated categories of things, namely, true fossils and minerals, or to agree that the true fossils were the remains of once living creatures. The fossilists collected them, described them, and illustrated them (often very accurately), and then struggled to explain them in terms concordant with Genesis.

Genesis is very clear on what went into the ark with Noah and his family:

> *They and every beast after his kind, and all the cattle after their kind, and every creeping thing that creepeth upon the earth after his kind, and every fowl after his kind, every bird of every sort.*
>
> *And they went unto Noah into the ark, two and two of all flesh, wherein is the breath of life.*
>
> (Genesis 7, 14–15).

Nothing is said about marine life, but literal interpretation of the text leaves no possibility of any species being left to be destroyed. The inevitable conclusion is that fossils without living counterparts cannot be the remains of former life and must have some other origin.

Martin Lister (1639–1712), who did most of his scientific work while practicing medicine in York, described and illustrated hundreds of "testaceous bodies" from the sea around the British Isles and included many fossil shells he had collected far inland (Lister, 1678). The gross similarities between living and fossil shells were obvious to him, but he also recognized the more subtle differences in morphology and realized that they were just as important. On that basis he refused to accept an organic origin for his fossil shells and, as Mercati had done 100 years

FIGURE 4.4. A sea urchin being examined. (Scilla, 1670)

before, interpreted them as *lapides sui generis* (stones of their own kind). He and Hooke seem to have had some lively debates on that subject, and Lister may have been the main critic of Hooke's ideas on fossils, but neither convinced the other.

Their friend Robert Plot (1640–1696), another avid fossil collector, became the first keeper of the Ashmolean Museum in Oxford, and wrote

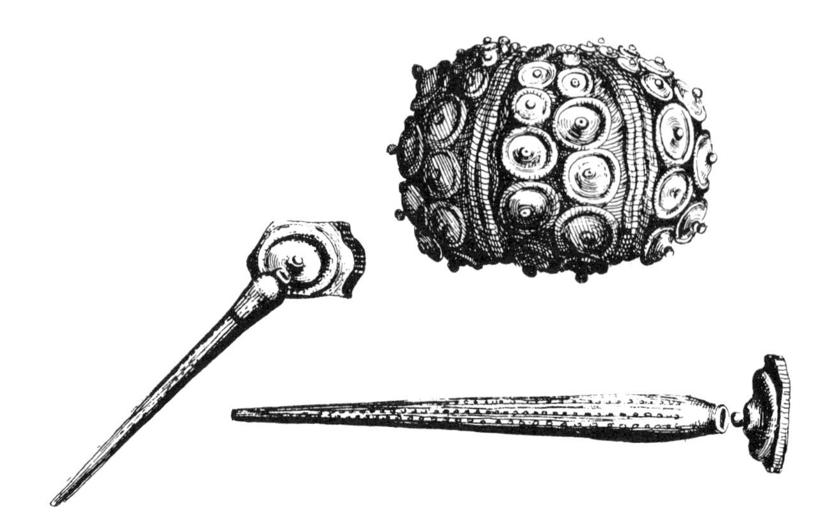

FIGURE 4.5. "Sea urchin, similar to those now taken from the seas that bathe Sicily, . . . stripped of its spines." (Scilla, 1670)

natural histories of Oxfordshire (Plot, 1677, 1705) and Staffordshire (Plot, 1686) that illustrate many fossils and some pseudofossils. He had no qualms about accepting Steno's definition of "chrystalls," but fossil bones and shells were another matter. He had an eye for the bizarre and a feeling for the mystical, which made it easier for him to take Lister's side in the controversy.

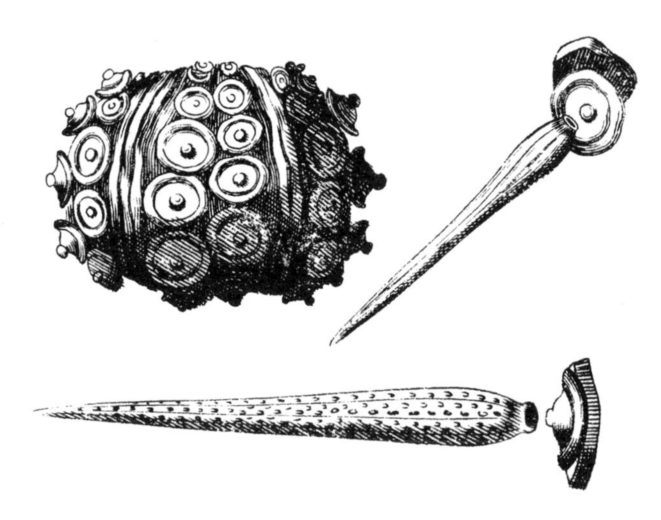

FIGURE 4.6. "A Sea-Urchin petrify'd with its prickles broken off, which are a sort of *Lapis Judaicus* . . ." (Ray, 1693). Illustrations were sometimes little changed from author to author.

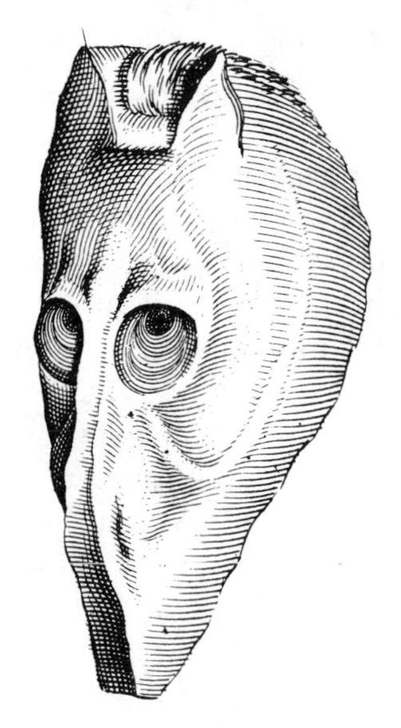

FIGURE 4.7. "Next we come to formed stones that resemble the parts of *four footed beasts*, whereof we meet with one sort in the Quarries at *Headington*, set in the body of the stone, the most like to the head of a Horse of any thing I can think of; having the *ears*, and the *crest* of the *mane* appearing between them, the places of the *eyes* suitably prominent, and the rest of the face entire . . . I . . . shall call them *Hippocephaloides*." (Plot, 1677, p. 127)

John Ray (1627–1705) was the son of a village blacksmith, but that did not prevent his capabilities from being noticed while he was still a schoolboy. He was sent to Cambridge, became a fellow of Trinity College in 1649, and was ordained in 1660. He had taught at Cambridge for about a dozen years when Episcopalian orthodoxy was reimposed in 1662, and all clergy were required to subscribe to the Act of Uniformity. He balked at the return to the old Anglican ritual, and thus was forced to withdraw from the priesthood and the College, and found himself unemployed. "Liberty is a sweet thing," he wrote to a friend (Ray, 1928, p. 25). Archbishop Ussher of Ireland, with his pronouncement that the

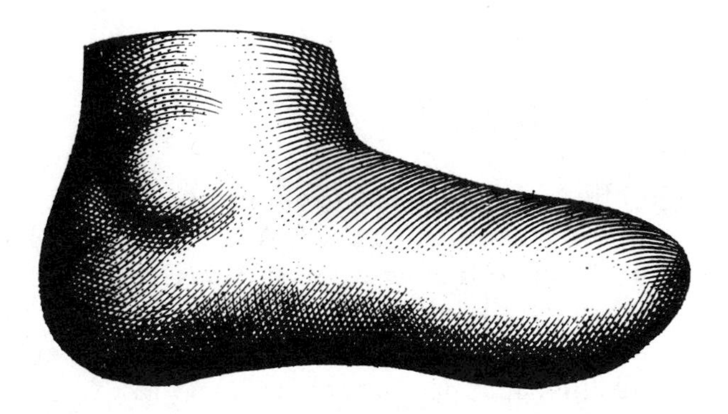

FIGURE 4.8. "Leg and Foot of a Man cut off above the *ancle*, which from the toe to the heel is about a yard long, and perhaps in the whole may weigh 50 or 60 pounds. But I take not this for a petrification . . . , but a stone formed in this shape purely by nature." (Plot, 1677, p. 139)

Earth was created at 9:00 A.M., October 26, 4004 B.C., was only an extreme example of the literalism of Genesis emanating from the English-speaking world.

Today it may be difficult to understand how very important religion was in British life well into the nineteenth century. The struggle between the Church of Rome and the Church of England had dominated the political scene for three centuries. Religious issues loomed large in Parliament and religious affiliation determined the succession to the throne. By the time the Act of Succession was passed in 1701, confirming the exclusion of Catholics from any consideration for high office, there had been enough splits in the Protestant ranks to keep religious controversies raging for another century and a half. At Cambridge and Oxford, professors and fellows were hired and fired for their religious views in spite of the growing independence of the Universities. They had been sacked wholesale when the Commonwealth swept in and the Puritans took over after 1649. Out they went again, by the dozens, after the Restoration of 1660. It was not so much a question of allegiance to one political group or another; it was a matter of principle. Englishmen felt strongly about the form as well as the substance of their religion.

Having taken his stand and lost his job as a result, John Ray in time became tutor to the Willughby family and from them derived a modest but steady income for the rest of his life. With Francis Willughby he made a tour of Europe and possibly met Steno in Montpellier, but after that he traveled very little; he enjoyed the quiet country life and devoted himself to studies of natural history. Ray made important contributions to botany and zoology, and even though geology never became his main interest, he had an excellent grasp of the controversy about the origin of fossils. His correspondence is exceptionally candid; most of it is preserved (Ray, 1718, 1846, 1928) and it illuminates the subject, as we shall see.

THEORIES OF THE EARTH

The educated public in Britain was becoming aware that great scientific advances were being made in the Royal Society and that this new knowledge would have an impact upon the classical conception of the world. Almost everybody knew Genesis, but now there was new interest in the "scientific" interpretation of the riddles it posed. A new book by a Cambridge divine, Thomas Burnet (1636–1715), was geared to this trend and was much talked about when it appeared, first in Latin (Burnet, 1681) and then in English translation in 1684 and 1689. The English version was called *The Sacred Theory of the Earth* and it promised to give "an Account of the Origine of the Earth, and of all the General

FIGURE 4.9. Frontispiece of Thomas Burnet's work attempting to reconcile an imperfect Earth with Genesis. (German edition, 1698)

Changes which it hath already undergone, or is to undergo till the Consummation of all Things." It includes long chapters "concerning the Primitive Earth and Paradise," "the Deluge and Dissolution of the Earth," and "the Conflagration."

Burnet argued that the paradisical world was perfectly smooth and spherical. It was composed of differentiated parts: a core of heavy particles surrounded by shells of liquid and air. The liquid differentiated into fat and oily liquids above and water below. Dust from the primeval atmosphere settled into the oily shell and eventually it became firm and habitable land. The Deluge resulted when this solid shell collapsed into the waters below and they gushed out to flood the Earth, leaving it in ruins. It is written in a rich, flowing style and still makes good reading, but many of the ideas it presents with such smooth assurance were debatable at the time and produced a strong response from theologians and scientists alike. It was no coincidence that the *Theory* was reprinted often and remained a best seller for decades.

Without mentioning Burnet by name, Herbert Croft, Lord Bishop of

Hereford (1603–1691) concluded "for certain that his man hath been to the Moon, where his head hath been intoxicated with circulating the Earth, and is now come down to us with these rare Inventions" (Croft, 1685, Preface). He accuses Burnet of "deducing one Errour from another" and being "besotted with his own vain and heathenish Opinions" (Croft, 1685, p. 178). It seems that Burnet had irritated the spirited Bishop by straying too far from the literal interpretation of Genesis and ascribing too much to the processes of nature and not enough to the direct action of God. "This savours very much of the Epicurean Opinion, who thought it below the Dignity of the Godhead to trouble itself with the minute Affairs of this lower world. But Believers know it is no trouble at all to God to act in infinite things, more than in one" (Croft, 1685, Preface).

A country parson in Suffolk, Erasmus Warren (?–1718), also attacked Burnet on theological grounds, without particularly illuminating the subject under discussion, but he called his book *Geologia* (Warren, 1690) and thus scored some sort of first in the use of that word. [Lovell (1661) had previously used the word as a heading, but only for a chapter on earths, and Escholt in 1657 had written *Geologia Norvegica*.] Warren protested that the rugged continents are beautiful to behold and not ugly ruins. Warren's public polemic with Burnet consists of a series of blasts and counterblasts: Warren (1690), Burnet (1690), Warren (1691), Burnet (1691). They "danced the same steps with greater exasperation" (Porter, 1977).

William Whiston (1667–1752), Newton's protégé and later his successor as Lucasian Professor of Mathematics at Cambridge, did not attack Burnet personally, but wrote his own *Theory of the Earth* (Whiston, 1696). It opens with a recitation of Newton's laws of motion, is cast in a format superficially resembling the *Principia*, and is illustrated with mathematical-looking plates, but the substance is mostly theological and Whiston is the straight fundamentalist. Incidentally, Whiston's succession to the Lucasian chair had been arranged by Newton, but later they had an altercation over biblical chronology and when Whiston was fired by Cambridge in 1710 because of some unorthodox religious opinions he had expressed, Newton did nothing to help him.

Another Newtonian disciple and later the Savilian Professor of Astronomy at Oxford, John Keill (1671–1721), attacked Burnet's *Theory* and Whiston's counter-theory both at once (Keill, 1698, 1699), taking an even more conservative position. He was a High Churchman and the sum of his argument is "an evident demonstration of the impossibility of all *Natural* and *Mechanical* explications of the deluge whatsoever" (Keill, 1698, p. 30). To him it is "both the easiest and the safest way, to refer the wonderful destruction of the old world to the Omnipotent hand of God, who can do whatsoever he pleases" (Keill, 1698, p. 33).

An avowedly nontheological critique came from John Beaumont, Jr. (?–1731), a virtuoso who lived in Somerset and had a fine fossil collection (Beaumont, 1693). He goes through Burnet's theory section by section, pointing out errors in geology and physics, many of them fatal to Burnet's arguments, but then he becomes involved in elaborate interpretations and interpolations of ancient writings, without much benefit to his position. John Ray had been anxious to read his book, having heard of it a few years before it appeared (probably through Lhwyd and Lister; see Lhwyd 1945, p. 139). When he finally got to see it, he wrote to Lhwyd: "I have now read Mr. Beaumont's Considerations upon Dr. Burnet's Theory of ye Earth; & doe think that he hath fundamentally overthrown it . . . But a great deal of stuffe he hath about the mysticall and Allegorical Physiology of the Antients, wch I understand not, nor I believe himself neither" (Ray, 1928, p. 242). (For an account of the substantive issues involved in the battle between the theorists, see Nicolson, 1959.)

NATURAL THEOLOGY

Ray was unimpressed with Burnet, either as a theologian or as a naturalist, and he considered the *Theory* "no more or better then a meer chimaera or Romance" (Ray, 1928, p. 237). As much as he must have felt tempted, it would not have been Ray's style to attack Burnet personally in print. He gave vent to his opinion of the *Theory* in private letters, but kept his peace in public. Still, he was comfortable with the quality of his own theology and would not mind displaying some of it to the reading public. He knew the publishing business, was well aware of the commotion the *Theory* had caused, and must have had a good idea of the market potential for that kind of literature.

He dusted off some sermons he had given in Trinity some 30 years before, spruced them up, and sent them to Samuel Smith, the publisher in St. Paul's. Smith must have had his doubts because he printed only 500 copies according to Ray's contemporary biographer William Derham; his fears were not warranted, however, and *The Wisdom of God Manifested in the Works of the Creation* (Ray, 1691) sold out quickly. A much-enlarged second edition was out within a year, presumably in a much larger printing. Before long that was gone too and a third edition appeared in 1701. What had begun as an overt act of piety was now a literary hit and, at the same time, a source of much-needed income.

With one good seller in hand, Smith could have been expected to encourage Ray to write another popular title, and how could he have refused? Without delay he produced the *Miscellaneous Discourses Concerning the Dissolution and Changes of the World* (Ray, 1692), again

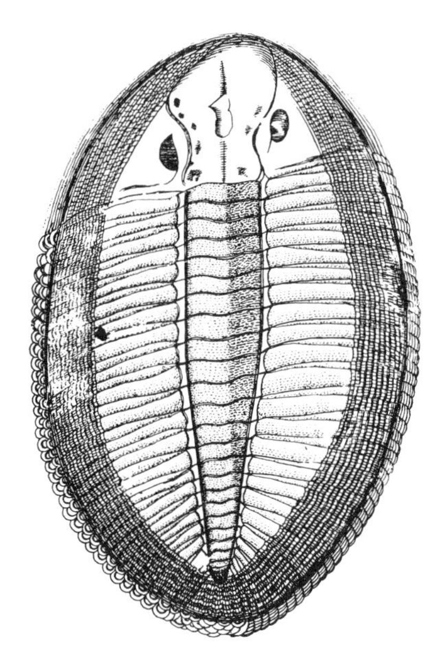

FIGURE 4.10. "[This] we found near *Lhan Deilo* in *Caermarthenshire* . . . whereof we found great plenty, must doubtless be referred to the Sceleton of some flat Fish." (Lhwyd, 1698)

based on a sermon he had given in Cambridge long ago. Again it sold out and again a second and enlarged edition appeared in 1693, a third in 1713, a fourth in 1721, and a "fourth edition corrected" in 1732. (Beginning with the second edition, the title was changed to *Three Physico-Theological Discourses.*) In his own words, he was "huddling up and tumbling out Books" (Ray, 1692, Preface). They were popular theological books, but in them he presented his geological ideas.

Ray had seen fossils in collections and in the field when he traveled on the Continent with Francis Willughby in the 1660s, but when he returned home he went to work on plants and animals, which is where he earned his reputation. He became Britain's leading botanist and in that capacity he received a letter in June 1689, from a young Welshman who was then the low-paid assistant of Robert Plot in the Ashmolean museum in Oxford and later became his successor there (Ray, 1928, p. 186; Lhwyd, 1945, pp. 87–88). His name was Edward Lhwyd (1660–1709) and he had a way of making friends by mail. In his letter he had included a catalog and specimens of some plants from Wales, including a few that had not been described before. Ray was grateful for such an addition to his *Catalog of English Plants* (Ray, 1670 &c.) and was properly impressed by Lhwyd's ability as a botanist. Actually, Lhwyd was then beginning to classify the Ashmolean collection of fossils. As their correspondence developed, he repeatedly brought up the subject of fossils, and Ray's interest in geology was rekindled. Ray and Lhwyd never met face to face, but they enjoyed each other's freely expressed opinions delivered by the post.

Ray's clear perception of what could and could not be expected in nature was important for Lhwyd as an antidote to the influence of Plot, who tended toward fanciful interpretations. In an era where systematic study of nature was just beginning, the young naturalist never could be quite sure of what to believe. Almost everybody still agreed that witches and hobgoblins were real, and seemingly intelligent people reported outrageous fantasies as if they were fact.

Ray was a polite but shrewd judge of human nature and he did not mind letting his young friend know where he stood. "And for Father Kircher I account him a credulous person, possessed of ye vanity of most of ye Religious of his church who delight to tell strange and miraculous tales to amuse and delude ye vulgar," he wrote to Lhwyd in 1690 (Ray, 1928, p. 211). "I never yet saw nor hope to see any repraesentations of Men, Women, Beasts, or Birds in marble or other stones," he wrote, but when Lhwyd related a new version of that ancient myth of a live toad being found inside a solid stone broken by quarrymen, this time reported by a Dr. Richardson from Yorkshire, Ray did not deny it, but could "not be perswaded that that Animal was at first spontaneously generated there" (Ray, 1928, p. 227).

Toads in Stone—A Recurring Tale

Edward Lhwyd to John Ray

Oxford, Feb. 30, 1691 (/2)

[Dr. Richard Richardson of North Bierley in Yorkshire] informs me that he was present when a stone was broken by workmen, which lay upon the top of the ground, wherein was contained a toad, in form and colour altogether resembling the common one, though something less, which, being laid upon the ground, crawled about as long as the sun shone warm upon it, but towards night died. I examined the stone (says he), and supposed it at first to be of an extraordinary open texture, or else the hole wherein the toad lay to have some private communication with the air; but upon a more strict inquiry I found the stone of a close grit, but that place especially where she lodged to be of a much harder texture, much of the nature of the iron stone which the workmen call an iron band (Gunther, 1945, p. 157).

John Ray to Edward Lhwyd

Black Notley, April 5, 1692

Richardson. Yet can I not be perswaded that the Animal was at first spontaneously generated there, or that so soft a creature could possibly make it self a hole in so hard a stone; but that when it first crept in there ye stone was soft and yielding, & afterwards hardened about it; & doubtlesse some air must some way or other insinuate it self into ye stone for its respiration. (Ray, 1928, p. 227)

At Passy, near Paris, April 6, 1782, being with M. de Chaumont, viewing his Quarry, he mention'd to me, that the workmen had found a living Toad shut up in the Stone. On questioning one of them, he told us, they had found four in different Cells which had no Communication; that they were very lively and active when set at Liberty; that there was in each Cell some loose, soft, yellowish Earth, which appeared to be very moist. We asked, if he could show us the Parts of the Stone that form'd the Cells. He said, No; for they were thrown among the rest of what was dug out and he knew not where to find them. We asked if there appear'd any Opening by which the Animal could enter. He said, No. We asked, if, in the Course of his Business as a Labourer in Quarries, he had often met with the like. He said, Never before.

. . . The Part of the Rock where they were found, is at least fifteen feet below its Surface, and is a kind of Limestone. A part of it is filled with ancient Sea-Shells, and other marine Substances. If these Animals have remain'd in that Confinement since the Formation of the Rock, they are probably some Thousands of years old. . . .

B. Franklin

(Franklin, 1906, v. 8, p. 417)

These beds are void of Petrifactions, but on breaking a block in Bolsover Field, in the year 1795, of a ton and a half weight, a toad was discovered alive in the centre, which died immediately; no crack or joint was perceptible.
(Watson, 1811, p. 2)

The Workmen employed some years ago in getting this Marble from the Quarry, on Cowden, near Ashford, on breaking a solid block, found in the midst thereof two Toads alive, at the distance of about six inches of each other, which died immediately upon exposure to the air.
(Watson, 1811, p. 44)

Geological Wonder. An English paper states that the miners in Ridgehill coal pit, near Oldham, a few weeks since struck upon a rock, on cutting which, they found imbedded in a solid mass, a *frog alive!*—It was discovered at a distance of one hundred and four yards below the surface, and was of a coal color, but on being brought out of the pit, it became of the usual hue. It was alive some days after it was exhumed.
(*Boston Weekly Magazine*, v. 1, p. 211, March 9, 1839)

Animals Found Imbedded in Stone. Frequent instances, as it is well known, are recorded by naturalists, of toads and some other animals which have been found completely imbedded in solid rock. Some people are disposed to be somewhat skeptical in relation to these facts, but they are too well authenticated to admit of doubt, since there is nothing in them that savors of the miraculous. . . . Shell fish and toads, if imbedded in sand or clay, or any soft substance that should harden into stone, would lie there in a torpid state for ages. They are cold-blooded, and as long as no heat comes to them from without, they cannot perish. There is nothing contradictory to the laws of nature in these facts.
(*Boston Weekly Magazine*, v. 3, p. 278, May 15, 1841)

Ray was ready to agree that the slates from the coal pits contain the remains of real plants (Ray, 1928, p. 259 &c.), but familiar as he was with British botany, he could not recognize the species. He respected Scilla's view (1670) that fossils are the remains of real animals (Ray, 1928, pp. 265–266) but was reluctant to accept it because he could not conceive of a mechanism that could deposit shells in the mountains far from the sea and still be "reconciled to Scripture or Reason" (Ray, 1718, p. 257). Finally, in one of his last letters to Lhwyd in 1703, he wrote: "Only ye beds of Oyster-shels wch are found in Kent, Surrey & other places doe a little stagger me, so that all their circumstances considered I can hardly shake off my former opinion, that these were originally Beds of living Oysters, breeding and feeding in the places where they are now found, wch were anciently ye bottome of the sea" (Ray, 1928, p. 284).

LITHOPHYLACII BRITANNICI

Almost from the beginning of their collaboration by mail, Ray encouraged Lhwyd to publish the catalog of fossils he was compiling. By 1697 it was finished, but unlike his mentor, Lhwyd did not have the knack of dealing with publishers. He called it *Lithophylacii Britannici Ichnographia*, presented it to Oxford University for publication, and they turned it down. The commercial publishers in London would not touch it either, and their reluctance is understandable: The text is in Latin and the title in Latinized Greek. It contains a bare-bones list of the Ashmolean fossils, illustrated with a lot of figures that were not particularly well drawn. The only part that could have been interesting to a general reader, the six "letters" that are now included, had not yet been written when the manuscript was circulated to the publishers. Quite a few fossilists realized that this was to be the first illustrated scientifically classified list of British fossils, and that it would be of great use to them, but that failed to impress the booksellers. All they could see was great expense with the plates and very small sales.

The matter was resolved one summer day in 1698 when a group of Royal Society virtuosi met in a tavern and decided to publish the book themselves. Ten of them, including Lister, Newton, and Sir Hans Sloane (1660–1753) took 10 copies each, 20 were to go to the author, and that was that. "So we shall make our books worth double the value & be obliged to or putt on by no body," wrote Sloane (Lhwyd, 1945, p. 23) and he was right. About 30 years later, when the Woodwardian Library was being sold at auction, scientific books were bringing a few shillings each, but a copy of the *Lithophylacii* (Lhwyd, 1699) went for 22 (V. Eyles, 1971, p. 417).

Lhwyd's catalog was a milestone. Its 23 plates may be artistically undistinguished and a bit crudely engraved, but they illustrate the organic and the inorganic well enough. The classification makes very good sense in the light of the botany and zoology of the time, and provenance is clearly described for each specimen. Five of the six letters appended to the catalog discuss correlations of fossil and living forms and represent the first serious attempt at such a subject. Lhwyd identified the fossil remains of fish and echinoderms, noted the great variety of fossil cephalopod shells compared with the rarity of living Nautili, and mentioned the virtual absence of modern equivalents of the fossil flora of the Coal Measures. Only the belemnites confused him completely and he never could determine what kind of animal they might have come from.

The sixth letter, addressed to John Ray, presents Lhwyd's views on the origin of fossils. Like the rest of the book, it is in Latin, but an English version exists (Lhwyd, 1945, pp. 381–396). It appeared in the third and fourth editions of the *Discourses* (Ray, 1713, 1721, 1732) and thus became available to a wide audience. Lhwyd argues against attempts to explain fossils as products of the Deluge and proposes a hypothesis that harks back to the pseudo-Aristotelian idea of fossils growing in the rocks. He suggests that fossils grew where they are now found, from natural spawn transported there over great distances by "Exhalations which are raised out of the Sea, and falling down in Rains, Fogs, &c.," but then he produces such a powerful array of clear and cogent objections to his own hypothesis, and answers them with such feeble defenses, that one may suspect that he may have privately entertained a more direct explanation.

Another systematic collector of British fossils was John Woodward (1665–1728), Professor of Physic at Gresham College who was widely renowned in his time as a super-diluvialist. In his first major book, *An Essay Toward a Natural History of the Earth* (Woodward, 1695), he proposed that in the Universal Deluge all rocks became comminuted with the waters and gravity ceased to act. When it returned, all solid matter settled out again according to the density of the individual particles, the heaviest at the bottom. All strata were made that way, and thus enclosed the remains of animals then living, according to their size. The idea appeared preposterous to many of Woodward's contemporaries, but it had popular appeal and the *Essay* went through three English editions (1695, 1702, 1723) and was translated into Latin, French, Italian, and German.

Woodward was contentious. Having proposed emetics as a cure for smallpox, he assaulted Richard Mead, a physician who favored purging, and they fought an impromptu duel. He lost, and Mead could have killed him had he wanted to, but was content to leave him sprawling on the ground with his sword out of reach (V. Eyles, 1971).

Ray detested Woodward and wrote to Lhwyd: "as for dr. Woodward's *Hypothesis*, if he had modestly propounded it as a plausible conjecture, it might have passed for such; but to goe about so magisterially, to impose it upon our belief, is too arrogant and usurping . . . His motion of gravity is ridiculous . . . as also that of Springs &c from ye Abysse, the heat or fire being between" (Ray, 1928, pp. 256–257).

Another critic, John Arbuthnot (1667–1730), compared the ideas of Steno and Woodward (Arbuthnot, 1697) and concluded that Steno's theory was burdened with fewer difficulties than Woodward's. In Italy, Antonio Vallisnieri (1661–1730) and Anton-Lazzaro Moro (1687–1764), and in Germany, Elias Camerarius (1673–1734) and Gottfried Wilhelm Leibniz (1646–1716) all had unkind words for the *Essay*, but amid all that clamor and indignation none of the critics took the trouble to note that, apart from Hooke's long-unpublished work, it was the first book by a Fellow of the Royal Society that clearly accepted the notion that fossils are the remains of former life.

Woodward's collection was enormous, and his work connected with it now constitutes his major contribution to geology. The *Brief Instructions for Making Observations in All Parts of the World* he had printed for his correspondents and collectors (Woodward, 1696) is a model of excellent museum practice, and the massive catalog of his English minerals and fossils (Woodward, 1729) contains the best classification system then available. Classification was still a major problem in his time, and his catalog is the fruit of his lifelong struggle with it. It became the standard reference on the classification of fossils and minerals in eighteenth-century Britain.

In his will Woodward directed that the collection described in his catalog go to the University of Cambridge, and the rest of his property, including many valuable antiquities, be sold at auction with the proceeds invested in land to maintain a lectureship there. The terms of the will specify the topics of the lectures, to be based on Woodward's own books, and one could take the view that the real purpose of the benefaction was not so much to serve geology at Cambridge, as to perpetuate Woodward's own views (Clark and Hughes, 1890, pp. 166–189; Gunther, 1937, pp. 425–430). Curiously enough, both aims were ultimately satisfied. The lectureship developed into the Woodwardian chair in Geology at Cambridge, and the collection is still there, perhaps the oldest geological collection preserved intact. It is now housed in the museum named after an illustrious incumbent of the chair, Adam Sedgwick.

In continental Europe, Woodward's fame was promoted by Johann Jacob Scheuchzer (1672–1733), teacher of mathematics, town physician, and zealous fossil collector in Zürich. Scheuchzer had translated Woodward's *Essay* into Latin and thus had made it available to a new audience. He was a product of the ultraconservative and scholastically oriented Swiss educational system—deeply religious, and scientifically impressionable. He had thought about fossils but had come to no conclusion about their origin until he read the *Essay*. Then he became totally converted to Woodward's views and spent his life expounding and defending the diluvialist position.

As Woodward's disciple he was elected Fellow of the Royal Society and became a frequent contributor to its *Transactions*. He shared in the widespread veneration of Isaac Newton, but his first major work, a textbook of physics and natural history (Scheuchzer, 1716), made clear that he, like many other vociferous admirers, also failed to grasp the physics of the *Principia*. His view of nature was still Aristotelian, and his book presents nothing that could have been new at the time, but it was a very successful publishing venture and went through five editions, establishing his reputation as a scientist.

In his long career Scheuchzer described hundreds of fossil plants and animals, and he thought of them all as remains of the Universal Deluge. When he found two fossil vertebrae in the Lias near the gallows of Altdorf (east of Nuremberg), he was convinced that they were human (Scheuchzer, 1708, p. 22 & pl. 3) even though they looked anything but that (see Jahn in Schneer, 1969, pp. 193–213, for an excellent account of Scheuchzer's pseudohuman fossils). His friend, Johann Jacob Baier (1677–1735), professor of medicine at Altdorf and a careful student of the fossils from that area, described similar specimens as fish bones (Baier, 1708), and so had Lhwyd (1699), but that made no difference to Scheuchzer.

As a physician, Scheuchzer would have been intimately familiar with the shape of human vertebrae. Baier even wrote him, after he had seen his assertions in print (October 7, 1708), pointing out that those vertebrae showed "no traces of that canal which provides the general opening for the passage of the spinal cord" (Baier, 1958, p. 93; Hölder, 1960, p. 365). He found it necessary to remind Scheuchzer that the location of the gallows would be no evidence for a human origin of these vertebrae. One of the specimens is still preserved in the paleontological museum of the University of Zürich, and it clearly came from the skeleton of one of those ichthyosaurs for which the German Lias is famous; but for Scheuchzer it always remained a specimen of fossil man.

HOMO DILUVII TESTIS

Scheuchzer's obsession with the search for remains of the miserable sinners who perished in the Flood eventually made him famous when he thought he had found articulated skeletons of them in the quarries of the Schiener Berg near Oeningen, on the German side of the Rhine, east of Schaffhausen (Scheuchzer, 1726*ab*). (The richly fossiliferous calcareous shale there is now known to be a late Miocene freshwater lake deposit.) In an excited letter to Sir Hans Sloane (Scheuchzer, 1726*a*) he

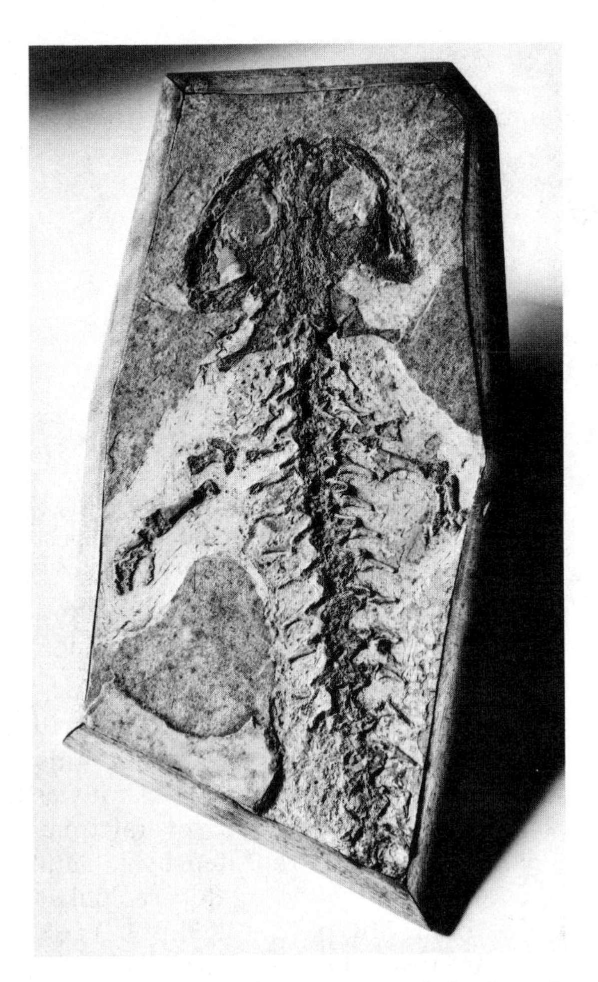

FIGURE 4.11. Scheuchzer's *Homo diluvii testis*, carefully cleaned up by Cuvier. (Photograph courtesy of the Teyler Museum, Haarlem)

reports that he identified more than enough skeletal elements in the skulls to prove that these were fossils of adult humans, *Homo diluvii testis*—the witness of the Deluge. That result was implanted in the scientific literature and propagated for a generation until it was put to rest by Cuvier (1825, v. 5, pp. 431–440) who cleaned Scheuchzer's best specimen (then and now in the Teyler Museum of Haarlem) and showed that it was no hominid but a large salamander. Curiously, the quarrymen of Oeningen always supplied Scheuchzer with skulls and foreparts of these fossils, but never with the other end, which would have included a long, massive tail.

A variety of fossil bones, collected here and there, had been described and identified as human, no matter what they looked like. Perhaps the liveliest such report comes from Battista Fregoso (or Fulgosus,

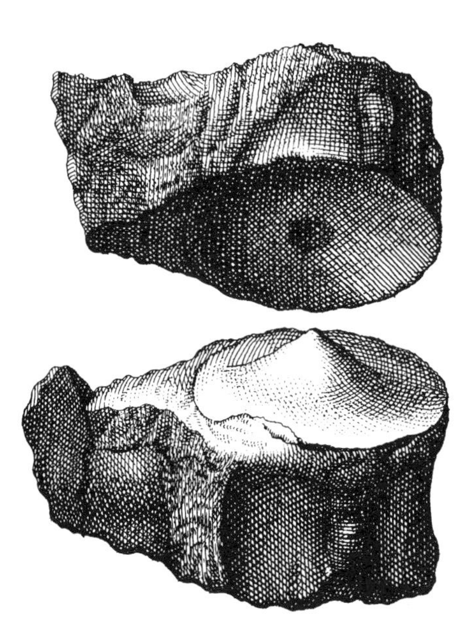

FIGURE 4.12. The vertebrae Scheuchzer regarded as human. One of them is still preserved in Zurich. (Scheuchzer, 1708)

1453–1504), a retired Doge of Genoa, who wrote that "many serious people" witnessed the discovery of a seagoing ship, with the bodies of 40 men in it, 100 fathoms underground in a metal mine in the Canton Bern in 1460. He had written the account in 1483; it was printed much later (1509, book I, chapter 6), reprinted several times, and became widely accepted as factual. It was repeated by Johann Jacob Wagner (1680, p. 356) and seriously discussed by Robert Hooke (1705, p. 438f). It was cited by de Maillet (1748, p. 79; 1968, p. 92) as evidence of the universal ocean and the great antiquity of man on Earth, and by Buffon (1749) to demonstrate the changeability of sea into land.

Carozzi (in de Maillet, 1968) has looked into the possible origins of that story and concludes that the mine was probably in residual iron ore

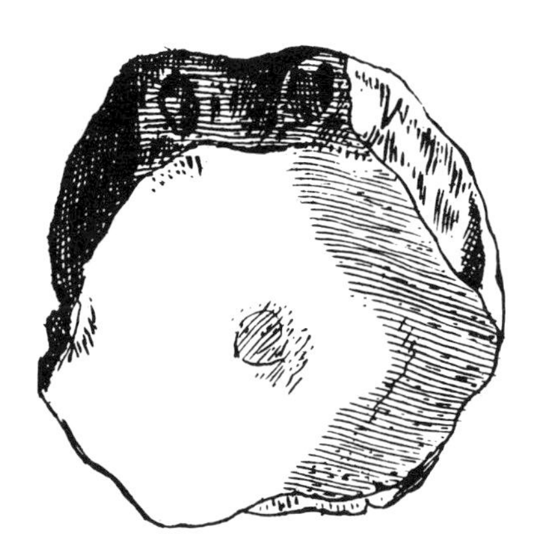

FIGURE 4.13. "Ichthyospondyli: Latrunculus subrotundus vulgatior, ex aestuario Sabriniano," or a roundish fish bone shaped like a checkerpiece, from the Severn estuary. (Lhwyd, 1699)

on the buried karstic surface of folded limestones, perhaps in the Delémont basin, about halfway between Bern and Basel. Overlying Tertiary sandstones contain fossil animal bones which are often found reworked with the iron ore in sinkholes, and fossil wood is known in the area. Fossil wood frequently has caused confusion. (Silicified logs found lying on the surface in Egypt and other places were often identified as masts of ancient ships.)

Within the straits of scriptural orthodoxy and in an intellectual climate still imbued with scholastic ideas, Scheuchzer's diluvialist paleontology was comfortably acceptable. He maintained a tremendous correspondence and was widely quoted—to the point that Moro (1740, p. 187) dubbed him "the Helvetic Pliny." By accepting Woodward's contrived blend of physical and supernatural causes for the Deluge, he effectively defused the arguments against the organic origin of fossils, and therein lies his greatest contribution. The impetus he gave to the study of European paleontology and paleobotany was felt for about a century.

One of the last to attempt an "impartial" discussion of the various hypotheses of the origin of fossils was Antonio Vallisnieri (1661–1730), physician in Reggio, later professor at Padua, and Fellow of the Royal Society since 1705. He was primarily a zoologist but made an important contribution to geology when he demonstrated that springs in the mountains, large and small, really do derive their water entirely from rainfall and melting snow without any need for the supposed underground connections with the sea or any mysterious process for raising the water and removing the salt from it (Vallisnieri, 1715). The ancient notion of large amounts of water surging through underground cavities was a fundamental ingredient in Woodward's theory of the Deluge. Having effectively exploded that mechanism, Vallisnieri now found himself without a viable vehicle for the organic origin of fossils and was obliged to reject it, together with all the other hypotheses that had been proposed (Vallisnieri, 1721). Having recognized the cardinal flaws in Woodward's (and Scheuchzer's) thinking about the Deluge, he could not have taken any other position in public. His inner thoughts on the subject do not appear to be recorded, but it is known that he considered unorthodox solutions to other problems (Vallisnieri, 1905). He realized that the differences between species are not abrupt, that the "chain of being" or progression and connection of all living things has real meaning insofar as plants and animals are concerned; and he entertained the highly heterodox idea that all living things—not just humans—must have souls.

Fanciful and supernatural explanations of the origin of fossils were common in the early eighteenth century until a comical incident put a halt to such speculations. Dr. Johann Bartholomaeus Adam Beringer (1667–1740) was dean of the medical school of the University of Würz-

FIGURE 4.14. One of Beringer's "lying stones," carved with what the pranksters thought might be taken for Hebrew writing, but which no Hebrew scholar we know has been able to decipher. (Beringer, 1726)

burg, court physician to the Prince-Bishop, and an ambitious virtuoso. He must have seriously offended two of his colleagues, a librarian and a professor of mathematics, because they conspired against him in an original way. They fabricated some outlandish petrifactions from a soft local limestone, hired a farm boy, and sent him to Beringer with the story that he had found the "fossils" on a hill just out of town. Beringer was delighted, and the conspirators made more and more specimens, sent them to him, and even let him find some himself on that hill. It was a huge joke until one day the conspirators learned that Beringer had hired an engraver and was preparing to publish his "discoveries."

They tried to convince Beringer that the stones were fake, even showed him how they made them, but he would not believe any of it. The Prince-Bishop failed to see the humor in making his physician appear a fool, a regular tempest erupted in the Würzburg teapot, but the book appeared, illustrated with 21 plates of hilarious pseudofossils (Beringer, 1726), a monument to the unbridled imagination of the conspirators. More important, it was a warning to other virtuosi to be a little more careful. (For a full story of the "lying stones," see Jahn and Woolf, in Beringer, 1963.)

Last in this parade of fossilists is a retiring Italian priest, Anton-Lazzaro Moro (1687–1764), who was for many years director of the seminary at Feltre in the foothills north of Padua, and then served the cathedral of Portogruaro east of Venice. He was born and died in the small town of San Vito al Tagliamento, traveled very little, and was never elected Fellow of the Royal Society. He lived a quiet life but wrote a very important book *De Crostacei e Degli Altri Marini Corpi . . .* , (Moro, 1740) in which he effectively summarized the problem of marine fossils, showed a profound understanding of the dynamic history of the Earth, and turned geology in new directions. He had the pleasure of seeing his book abstracted in the *Philosophical Transactions* and his ideas debated among the virtuosi.

Moro was an ultraplutonist. He was awed by accounts of the eruption of a new island out of the sea near Santorin in the Aegean in 1707 (Vallisnieri, 1721), and the similar birth of Monte Nuovo on a plain near Pozzuoli, west of Naples, in 1538. He came to believe that all mountains, islands, and whole continents were produced that way by the Earth's internal fire. The first half of his book (Moro, 1740) is a polemic against Burnet's and Woodward's pseudoscientific analyses of the Noachian Deluge, building on the arguments of Vallisnieri (1721). "The Deluge ought to be believed according to the Scripture, as a miracle, and not to be proved by natural rules," writes Father Moro (*Phil. Trans.* v. 44, p. 164).

The summary of his own "system" begins with an orthodox account of the Creation, but then proceeds without any further need for divine action. There is providence and a Divine Plan, but no more supernatural effects. Life develops in favorable environments, plants first and animals after them, in the sea and then on land, and man appears as the ultimate product of that development. "Marine animals and plants," writes Moro, "the spoils or remains of which are still found today on or under certain mountains, were born, nourished, and grew in marine waters until those mountains were raised above the surface of the sea and then were extinguished, and furthermore continued to exist in a petrified state while those mountains, having left the water-covered bosom of the Earth, attained those heights in which we now see them" (Moro, 1740, pp. 425–426). The cause of the upheavals that submerge the land or raise the bottom of the sea above the level of the waters is the internal fire of the Earth. All rocks, even limestone, are derived from volcanic vents, and new creatures develop in progressively higher strata, according to the nature of the environment. Mountains of unstratified rock he declared to be igneous and classified them as *Primary*. All other mountains, consisting of layered rocks, he called *Secondary*, and he was the first to make that distinction.

He sprinkled the book with affirmative references to Genesis, but even so one may be surprised that he was able to steer such a collection of revolutionary concepts past the censors in Venice. He must have had some good friends in important places. There was an uproar in clerical circles when the book was published, and for a time it was difficult to find a copy, but its fame soon spread, two more editions appeared in Germany (in German; 1751, 1765), and Moro's position was established.

The division into Primary and Secondary rocks was further expanded by another perspicacious Italian, Giovanni Arduino (1714–1795). He was a citizen of Venice, a prominent mining engineer, and an unusually observant student of the neighboring areas of the Alps. He published little, but made his unpublished work available to the Swedish mineralogist Johann Jacob Ferber (1743–1790) when Ferber visited Venice in 1771, thus getting at least an abstract of his ideas into print (Ferber, 1776).

Arduino's *Montes primarii,* the primary mountains, consist of contorted "mica-slates," perhaps resting on granite in some places. They are the ore-bearing mountains with metallic veins. The "slates" (schists, in modern terminology) are overlain by dense, fine-grained limestones, the "calcareous Alps" or *Montes secundarii,* "containing now and then some marine petrifactions" (Ferber, 1776, p. 39). Next come the "Lower Hills" or *Montes tertiarii,* lying in valleys or on the slopes of the calcareous Alps, "partly produced by their decays and accumulated sand and clay-beds." These formations "have also suffered many devastations by volcanic eruptions," but "some of these *Montes tertiarii* are posterior to the eruptions" (Ferber, 1776, p. 46). In general "they contain plenty of petrifactions" but, judging by the fossils Arduino mentioned by name, he obviously included in his "tertiary" a number of formations now known to have widely different ages, many much older than Tertiary.

CHAPTER FIVE

The French Connection

FROM CARDANO TO PALISSY

The French are not like the British, and *vivé la différence.* Geology came of age in Britain, and the historians of British geology tend to be smug about it. Occasionally they forget that some of the ideas most essential for the maturing of geology had come from France.

Beginning with Bernard Palissy (?1510–1590), the potter-turned-philosopher, a French geological literature with an approach of its own developed. Theological pressures played a surprisingly small part in it. Not that France did not have its share of religious warfare—on the contrary. The persecution of Huguenots was every bit as savage as the burnings in Italy or the beheadings in the Tower of London.

By nature, Palissy was a questioner and a doubter. His inquiries in religion led him to become a Calvinist at the age of 36, and as a result he narrowly missed the gallows in Bordeaux 13 years later. Somehow he escaped the St. Bartholomew's Day Massacre in 1572, but in 1588, at the age of 78, he was imprisoned again for reasons that are not now altogether clear, and he died in the Bastille two years later. The French difference was that he was being persecuted mainly for his religion, not because of his scientific books and lectures.

Palissy was a painter, a stained glass artist, a potter, and a surveyor, which means that he must have had some training in mathematics. He became famous through his "figulines," a tin-enameled pottery heavily decorated with naturalistic motifs, including casts of real leaves, shells, crayfish, and the like. He moved from Saintes (on the Charente, north of Bordeaux) to Paris in 1567, branched out as a designer and decorator, and became widely known. In 1563, encouraged by the prospect of religious peace after the Edict of Amboise was signed, Palissy had published a remarkable book called *Recepte Véritable,* giving advice on agriculture, gardening, irrigation, natural and religious history, and philosophy for good country living under the benefits of peace and the Reformed Church (Palissy, 1563). He was not convinced that the peace would endure for long and he ends the book with a promise to design an impregnable fortified city where the Protestants could be safe from attack. His plan was based on the observations he had made of the way snails build their shells.

Palissy was not an academic type. On the contrary, he made it a point to present himself as the unlettered artisan, unbound by Scholastic dogma, but as he grew older he felt the urge to communicate his experiences and his philosophy to the intellectual public. In 1575, with religious freedom again reestablished, he advertised and gave a series of lectures by subscription. They were very successful and continued for about a decade. What the academics of Paris thought of the lectures is not recorded, but they must have been horrified because Palissy's views often ran counter to what was then considered established knowledge. Furthermore, he had a teaching collection and brought specimens into the lecture hall to demonstrate his points! One can only guess that the lectures developed from the ideas of his first book and that their substance is covered by his second book, *Discours Admirables* (Palissy, 1580).

Judging from that book, Palissy was an original thinker, a dedicated experimenter, and an accurate observer of nature. He had a clear understanding of how water moves on the surface and in the ground and he must have had experience in irrigation. He recounts the elaborate tests he made to develop his pottery, and apparently he grew crystals and made chemical preparations. When he was asked about the fossils in the hills, he went out to look for them and must have done a fair amount of what now would be called field geology.

Conventional wisdom of Palissy's time accepted the ancient view that water from the ocean travels into the mountains through underground passages, losing its saltiness on the way, then issues from springs, and ultimately finds its way back into the sea. "All rivers run into the sea; yet the sea is not full; unto the place from whence the rivers come, thither they return again," says the Bible (Ecclesiastes, I:7), and that view was maintained in great Scholastic style by the learned and influential Athanasius Kircher (1602–1680), that "credulous person," a century after Palissy. The analogy with the circulation of blood in the human body, through veins and arteries, played a part in this concept (Kircher, 1665, v. 1, p. 240), together with the observations that springs continue to flow in places that "never" receive any rain, as in the mountain monasteries of Egyptian deserts.

Palissy states categorically that all surface water comes from rain. Some water sinks into the ground and flows through fractures and through porous rocks, such as sandstone. It cannot flow through

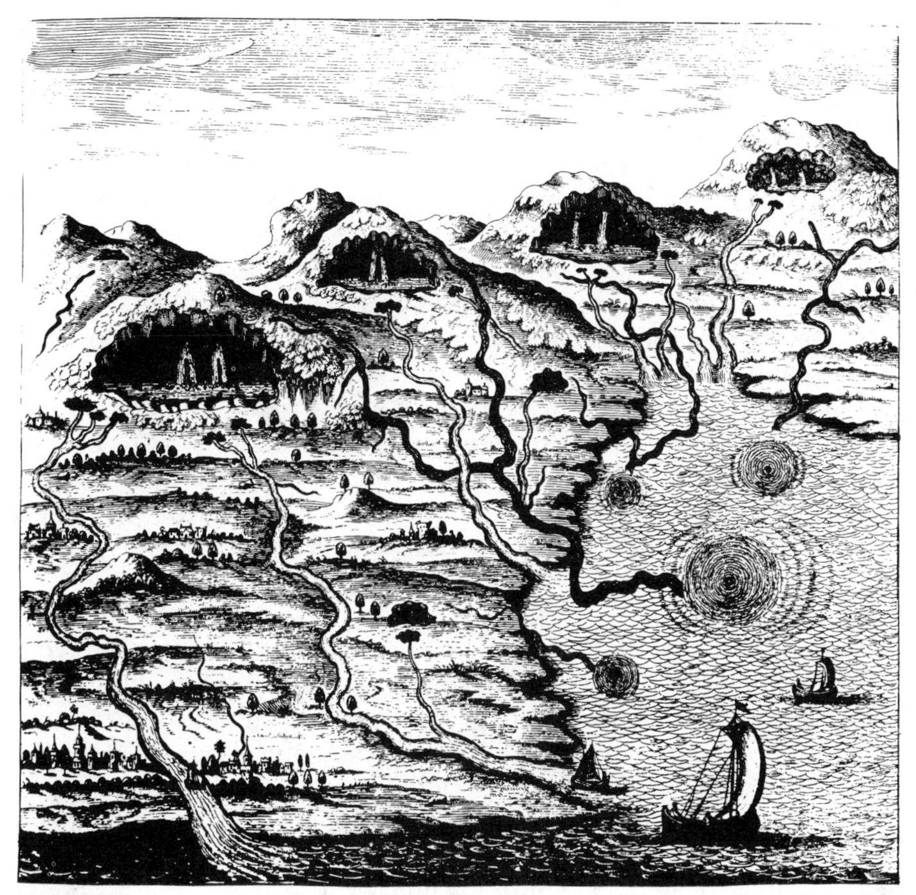

FIGURE 5.1. "Just as blood circulates in the human body, water rises from the sea through underground veins into subterranean fountains from which it issues in springs, losing its saltiness in the process." (Kircher, 1665)

bedrock or through impervious layers, such as clay. The underground flow is very slow, which explains why springs continue to flow long after it stopped raining. "The reason then why waters are found either in springs or in wells, is none other than that they have found a rocky bottom, or one of earth and clay which can hold water as well as the rock; and if someone seeks water in sandy ground he will never find any unless there is underneath some clayey earth . . . " he writes (Palissy, 1957, p. 56).

Mountains are there, wrote Palissy, because they are stronger than the plains and resist being washed away by water. Frost and water make rocks crumble and if rock had been made only at the beginning of the world, none would be left by now. Rock does not grow (like a vegetable) but is generated by congelation "in deep and hidden places" (Palissy, 1957, pp. 147–148). Rocks and metals are deposited from salt-bearing waters, and crystals grow only in such solutions (Palissy, 1957, pp. 150, 165). Animals and vegetables can be "lapified" by salt-bearing solutions, but some of "the softer parts rot away before being petrified" (Palissy, 1957, p. 164). Fossils found in the hills are the remains of animals that once lived in those very places, and some of those animals are similar to present sea life, but others are altogether unlike the creatures now living in the sea.

Much has been written about the supposed originality of Palissy's ideas or the lack of it. (For a good review see La Rocque, 1969.) Some of his inspiration certainly came from Italy, mostly through Cardano, but to claim that he was a simple plagiarist would be absurd. For some reason Palissy picks a fight with Cardano in the *Discours* (Palissy, 1957, pp. 155–161) and undeservedly calls him a diluvialist (perhaps only to minimize his own indebtedness to him), but at the same time he does not conceal that he has read *De Subtilitate* (probably in the French translation of 1556), and that he generally agrees with Cardano's geological ideas.

In the *Discours* Palissy spells out a wide range of observations and experiments he made as part of his geological studies, and that work alone would have been sufficient to make him a major figure. Like every other scientist, he also had his share of ideas that did not work out, but that does not obscure his obvious accomplishment of having been the first to introduce observation and serious discussion of geological subjects into the mainstream of French culture.

He was ahead of his time and his influence took some time to establish. The collection he had kept in his house disappeared and his books became scarce; they were not reprinted until 1636 and 1777. Mentions of Palissy's work began to appear while he was still alive, but his impact in paleontology, mineralogy, and hydrology was not actually recognized until the beginning of the eighteenth century. The 1777 edition of his works, edited by Faujas de Saint-Fond, lists 34 citations between 1584 and 1776, including Antoine de Jussieu, Buffon, Réaumur, de Maillet, d'Holbach, d'Argenville, Rouelle, and Guettard. [A romantic revival of Palissy developed in Victorian England (Morley, 1852), but its scientific significance was small.]

THE FLAMING EARTH OF DESCARTES

Next in the line of brilliant Frenchmen who left their mark in geology is René Descartes (1596–1650), who would not have considered himself a geologist even secondarily. His thinking centered on the idea that if one could only disregard all the concepts that have accumulated (many of which are wrong), and start anew with nothing but true principles, one should be able to develop all of science—the right way. He is recognized today as an important mathematician and philosopher, but he applied his method boldly to physics, medicine, natural history, theology, and law, making contributions of varying importance to all of them. His impact on geology stems from his view that the Earth was once a star like the sun, only smaller. The star had spots on it, like sunspots, and those were the places where it began to cool. The spots gradually enlarged and grew into a solid crust, water began to condense, and oceans formed on the Earth.

Descartes presented the idea in obscure language, perhaps deliberately obscure, but accompanied by extremely clear illustrations showing cross-sections of the globe with an incandescent liquid core and a layered crust that is broken—thus implanting the notion that the Earth's outer layers are unstable and more than enough heat remains inside the Earth to supply any volcano. That was definitely not an accepted view at the time, when the heat of volcanoes was generally thought to come from some sort of subterranean combustion of "bitumen and sulfur." The buckling of the crust, Descartes thought, may "form rocky shores of the sea & many mountain ridges, some very high and some gentle, and also shoals in the sea" (*Principia Philosophiae,* part 4, par. 44).

Descartes' geological writings are mostly in part 4 of the *Principia* [Descartes 1644; Haldane and Ross, 1911, translate only the headings; the French translation, Descartes, 1681, is in Adam and Tannery, 1971, v. 9 (2)], and they amount to only a small fraction of his large literary output. He writes about waters, tides, metals, mines, earthquakes, and magnetism, and most of his views are presented in a manner that makes them appear conventional for his time. Many of his speculations go nowhere, but he keeps stressing his own deductive version of the scientific method and rational causes for such things as the action of the lodestone, which most of his contemporaries still viewed as magic effects based on occult "virtues." His approach to science is mechanistic, and in that context it is understandable that he would see analogies between the Earth engine and the human-body engine, such as the veins-and-arteries interpretation for the origin of rivers. He must have been aware of Palissy's hydrology, but apparently did not care for it.

Descartes was searching for laws of nature and did not concern himself with interpretation of Genesis. He was a Copernican, of course, and had learned well from Lucretius, but he considered himself to be a devout Catholic. He mentions neither fossils nor the Deluge and surely expected no trouble from the Church. As long as he lived in Holland, he was safely out of its reach, but his conscience was tender.

The news of Galileo's trial in 1633 shook him severely and changed his whole attitude toward publishing. He had no taste for dangerous displays of the courage of his convictions and chose to retreat (see Descartes, 1911, pp. 118–119, 127; Scott, 1952; Vrooman, 1970). A major treatise he called *Le Monde*, discussing the Earth, its origin, and its constitution, was almost ready for the press, but he dropped the project. Postponing several smaller works, he wrote and rewrote his *Discours* (Descartes, 1637), and one can only guess what changes he made before he finally let it go—without his name on the title page. One fact is certain: His position on the heliocentric system came out carefully beclouded. When he later returned to geological subjects (Descartes, 1644), he wrapped them in elaborate double-talk and ended the last part *(De Terra)* with a sweeping submission to the Church:

> *[Side note:] But I submit everything I wrote here to the authority of the Church.*
>
> *And no less aware of my own insignificance, I affirm nothing: but all this I submit at once to the authority of the Catholic Church and to the judgment (of those) wiser than I; and I do not want anyone to believe anything except as persuaded by their own evidence and their own invincible reasoning.*
>
> (Descartes, 1644, par. 207)

It did not work. The Church put his works on the Index, and there they remained for three centuries. This, of course, did not stop them from being widely read and having a profound influence on scientific thinking for generations.

TELLIAMED

The last of this diverse trio of pioneers of the French geological tradition was Benoît de Maillet (1565–1728), a Lorraine nobleman who held various diplomatic posts, including 16 years' service as French consul in Cairo (from 1692 to 1708). He was anxious to learn everything he could about Egypt and the surrounding region; he learned Arabic, traveled, and asked questions. It was neither easy nor safe to travel around the Middle East in his day, but his position gave him the necessary political contacts and his private fortune the means.

From Cairo he moved to a similar post in Livorno, and there he studied that same classic landscape which had inspired his Italian geological forebearers. He saw Italian collections and was impressed by the Italian geological literature, particularly by Scilla's fine drawings of fossils (1670; see Carozzi in de Maillet, 1968). Livorno was where the shark had been brought ashore in 1666, and de Maillet must have crossed Steno's trail many times, but he never mentions him and it seems likely that he never knew of his work (Carozzi, in de Maillet, 1968). The intellectual (and political) importance of Florence had waned since Steno's time. His body had been brought there in 1687 for burial in San Lorenzo, the church of the Medici family (Scherz, 1971, p. 130), but barely a generation later his name seems to have been forgotten in Florence.

After his return from Livorno, de Maillet was sent on a mission to North Africa and the eastern Mediterranean for some five years, and in 1720 he retired and lived the rest of his life in Marseilles. There he collected his wide-ranging notes and wrote two important books: *Description de l'Égypte* (1735), and *Telliamed* (1748). For some reason he thought that his manuscripts needed professional editing by a capable humanist and he seems to have spent some years trying to find one. Finally he made contact with an experienced editor and translator, the Abbé Jean Baptiste Le Mascrier (1697–1760), and engaged him for the job. The Abbé took his time, and ultimately brought out the *Description* in good order, but *Telliamed* was another matter.

Telliamed was de Maillet's life work. It was a great theory of the Earth, supported by wide reading and far-reaching field observations, cast in the form of a conversation between an "Indian philosopher" and a "French missionary" in Egypt. Behind its quasi-oriental front, *Telliamed* was new and radical. It ignores God and does not even pretend to accept Genesis as fact. It builds on the ideas of Descartes and postulates a world made in a diminishing sea. All rocks are laid down in the sea, all landforms are generated by the sea, and all life comes from the sea. Creatures of the sea shed their fins and scales and move ashore as the sea diminishes. Mermaids and mermen change into women and men—and the skeletal similarities between fish and mammals are clear proof of these transformations. Such changes require a lot of time, and in his original manuscripts de Maillet writes of "more than 2 billion years" (Carozzi in de Maillet, 1968, p. 46) required for the retreat of the sea.

The concept of a diminishing sea follows directly from Descartes' theory of a cooling Earth, but the idea goes back to the early Greeks. The Mediterranean region, and especially southern France, offers many opportunities for geological observations that would confirm such a theory. The steady growth of the Nile delta was easily determined before the Aswan dams were built, and had been recorded since Herodotus. The silting-up of many North African harbors that had been open in Roman times was well known, and the retreat of the sea was obvious in the region of the Rhône delta. In many regions of southern France, fossiliferous limestones dip gently to the south and invite the interpretation that they have been deposited by a retreating Mediterranean Sea.

Publishing *Telliamed* was not a simple matter, and the Abbé went through elaborate contortions to make it acceptable. He rearranged the conversations into six "days" to make the book superficially similar to

Fontenelle's *Plurality of Worlds,* which was then very popular. He wrote a coy dedication to Cyrano de Bergerac (1619–1665), whose fantastic *Voyages to the Moon and the Sun* was then still widely read, and supplied a long, apologetic, and slightly confused preface that clouded the most controversial ideas. He softened the references to the very long time periods needed, and tried, with some success, to make the book more readable by providing smooth transitions between the six conversations. Even so he did not dare show his name in the book and signed the title page with the initials J.A.G. for Jean Antoine Guer, a lawyer and a writer who seems to have had little or nothing else to do with the project (Carozzi in de Maillet, 1968).

It had taken a long time, but in the end it all worked out. *Telliamed* was published in Amsterdam in 1748, in Basel a year later, and in The Hague in 1755. An English translation appeared in London in 1750 and was reprinted in Baltimore in 1797. The Hague edition included a brief biography of de Maillet by Le Mascrier, still totally anonymous and, unfortunately, not altogether consistent. That is the only contemporary biography we have of him, inaccurate as it may be.

In Paris in 1721, de Maillet read a slender book published that year and signed only with the initials H. G. I. D. P. E. C. D. R., presenting a system of the world even more outrageous than his own. That was Henri Gautier (1660–1737), *Inspecteur des Ponts et Chaussées du Roi,* the King's civil engineer. Apparently misinterpreting Descartes' ideas of gravity and the results of some dubious experiments with the mercurial barometer, Gautier calculated that at a depth of 1195 fathoms below sea level the pull of gravity would go to zero and then turn negative as one continued further down. That meant that another 1195 fathoms down one would emerge from an internal sea, the mirror image of the external one, so that the Earth must be hollow and proportionally as light as a paper balloon. The two oceans, internal and external, would be connected at the poles, Gautier thought.

Over the years, these ideas found a surprising number of defenders and they continue to reappear in various guises. Nevil Maskelyne, the Astronomer Royal who measured the density of the Earth, found it necessary to scoff at "some naturalists who suppose the earth to be only a great hollow shell of matter" (Maskelyne, 1775, p. 533), and the "hollow Earth" was seriously discussed in the United States in the 1820s (C. R. Hall, 1934, p. 149). (For an extensive analysis of Gautier see Ellenberger, 1975–1977.)

On this bizarre framework, Gautier elaborates several new and important geological ideas. "Mountains diminish without cease and never grow," he writes (Gautier, 1721, par. 2). Water from rain erodes the landscape, streams carry the sediment into the sea, and there it is deposited "in the greatest depths, in regular layers" (Gautier, 1721, par. 8). The action of rivers is greatest in times of flood and the average amount of solids carried by the water is about 1/1700 by volume. The amount of erosion is exactly equal to the amount of sediment carried away. As erosion proceeds, continents become lighter and the seas heavier, and the sediments in the sea harden into rock. Equilibrium is reestablished as the bottom of the sea is raised by "central gravity" to form new mountains, and the eroded continent subsides into the sea. Earthquakes occur when the crust of the Earth breaks during such readjustments.

These ideas did not suit de Maillet's concept of a diminishing sea, and he dismissed them out of hand. "If the author . . . had been acquainted with what happens in the waters of the sea . . . " he writes, he would not have found the "need to create a system so unnatural as his" (de Maillet, 1968, p. 136). Apparently he was not aware that Gautier had spent his life designing bridges, canals, and locks in southern France and that he was intimately familiar with sedimentation and with the action of rivers. Gautier's reasonable estimate of 1 part solids carried in 1700 parts of water somehow became "17 per cent of their volume" in *Telliamed* (de Maillet, 1968, p. 134), a 300-fold exaggeration. De Maillet went to great lengths to refute the hollow Earth and the attendant fluvialist concepts with the usual result of thus indirectly advertising them, but without much effect. Wrapped as they were in absurd theory, Gautier's pragmatic ideas had little response in French geology.

THE *JARDIN DU ROI*

De Maillet had served the government of Louis XIV (reigned 1645–1715) who was called the Sun King for good reason. During his unusually long reign France became the greatest power on the Continent and a serious competitor for Britain, but the distribution of power and wealth was quite different in the two countries. Whether or not Louis XIV had ever actually said "I am the State," it was an apt description of the political and fiscal structure he established in France, largely at the expense of the provincial nobility. Commerce and industry were encouraged by the king's centralized bureaucracy, but the encouragement was tied to central control. There was no room in France for such major private entrepreneurs as Sir Thomas Gresham, for example.

The support of medicine and science was part of the development and also came largely from the royal coffers. In London, the Royal Society was a private association whose Fellows derived from it recognition and entertainment but no income. In Paris, the *Académie des Sciences* also had formal autonomy, but in effect it was a government agency whose members received substantial salaries from the royal treasury. The various kings of England viewed the Royal Society with more or less favor, but they never controlled it. The kings of France, however, felt free to command the Academicians to make whatever studies suited the purposes of the government.

A good example of the royal philosophy was the *Jardin du Roi*, a garden for the cultivation of medicinal botany, originally set up by Louis XIII on the southeastern outskirts of Paris near the Sorbonne, to assure the best possible supply for the royal pharmacist. The garden was staffed by important scientists on good salaries, and their duties were arranged to leave them ample time for research and teaching. They were free to widen the scope of their activities, and the *Jardin* gradually became the focal point of all French natural sciences. Most of the geological research done in France in the eighteenth and early nineteenth centuries received support from the *Jardin du Roi* and its successors.

Antoine de Jussieu (1686–1758), the oldest in a dynasty of French botanists and professor at the *Jardin du Roi* since 1710, described fossil ferns of the St. Étienne coal basin (Jussieu, 1719) and recognized, as Ray had done in England, that they had no counterparts in the present flora of France. He had seen ferns from the West Indies that looked similar and concluded that the coal plants must have floated in from some such foreign place on some current of the sea. To de Maillet this was just another proof of the great universal ocean (de Maillet, 1748, v. 1, p. 84; 1968, p. 280) and he made the most of it. Jussieu had cited Palissy and made it very clear that it was not the Noachian Deluge he had in mind.

Perhaps the most talented antidiluvialist of that generation was René-Antoine Ferchault de Réaumur (1683–1757), mathematician, physicist, metallurgist, zoologist, and a commanding figure in French science in his day. He published little in geology (Réaumur, 1722), but exercised a disproportionate personal influence on the development of French geological thinking. In the Paris basin and west-central France, the regions he knew best, strata are almost horizontal and the low hills are covered with greenery. Almost no rock is ever seen at the surface, so the superposition of formations is far from obvious. This nature of the landscape led him (and other early Parisian geologists) to a "horizontalist" interpretation of fossiliferous rocks, which adopts Palissy's view that the animals lived and died in those places, but goes one step further in assuming that the different formations found in various places were deposited in different parts of the same ocean at the same time. That interpretation was adopted and long upheld by Réaumur's assistant Jean Étienne Guettard (1715–1786), a brilliant geologist, and severely restrained him in his otherwise far-reaching development of the first geologic maps.

Georges-Louis Leclerc, later Comte de Buffon (1707–1788), became director of the *Jardin du Roi* in 1739 and greatly increased its space, its scope, and its holdings. He was a financier, industrialist, versatile naturalist, best-selling author, and a serious Newtonian. His goal was to develop an internally consistent system of nature that required no supernatural factors for its operation. He presented his system in a great popular series, the *Histoire Naturelle* (Buffon, 1749), which ultimately ran to 36 quarto volumes, appeared in innumerable editions, and was the main vehicle for the tremendous influence he exercised in science.

He was a great writer and he portrays all nature in a vivid, storytelling manner, stone by stone, river by river, bird by bird, and beast by beast, up to and including man. He was keen on "facts and observation," made many geological studies in the field and experiments in the laboratory, and cautiously but clearly denies the theological approach to nature. "The force of impulsion was certainly communicated to the planets by the hand of the Almighty when He gave motion to the Universe; but we ought, as much as possible, to abstain in physics from having recourse to supernatural causes" (Buffon, 1792, v. 1, p. 74).

The first volume (Buffon, 1749), published right after *Telliamed*, presents his theory of the Earth, beginning with the origin of the solar system. He postulates the birth of the system in a grazing collision of a comet with the sun and a subsequent formation of all planets and their satellites from the ejected material—all very Newtonian, up to a point. He realizes that the present orbits could not have been derived from such a collision, casts Newton to the winds, and completes the picture without him. As part of his campaign to exclude the supernatural he mounts vigorous attacks on the theories of Burnet, Whiston, and Woodward, and condemns Scheuchzer for being "desirous of blending physick with theology" (Buffon, 1792, v. 1, p. 147), but then he also takes quick jabs at Leibniz, Steno, and Ray, implying perhaps that he did not realize they were on his side. He chides Leibniz for thinking that there is great heat in the Earth, then apparently unaware that the idea had come from Descartes.

Buffon describes the regular layering of the limestones (now known to be Jurassic) which he observed on the western and southern margins of the Paris basin and especially around Montbard (northwest of Dijon), where he lived. Not having seen many sandstones in the area, he doubts that they would be similarly layered, but he thinks of sandstone *(grès)* as something akin to granite. He reports having heated pieces of shale in a reverberatory furnace to the point where the outside melted, he observes how they hardened, and he understands that such hard minerals (what we now call silicates) weather back into shale again under the influence of water. Sand, he thinks, is nothing but the granules of such molten rock.

Buffon points out that fossil shells are common in many parts of the world and distinguishes between littoral and pelagic animals. Ammonites are pelagic, and their living equivalents have not yet been found. He describes volcanoes and believes that they derive their heat from the combustion of "sulfur and bitumen." He was aware that large earthquakes may not be associated with any volcanism and explains that they are caused by the explosion of gases in underground cavities. He denies that they could have anything to do with building mountains. If nothing else, there is drama and excitement in his stories of the oceans, with great whirlpools that swallow up ships and large areas of perpetual calm "where the art of the mariner becomes useless, and where the becalmed voyager must remain until death relieves him from the horrors of despair" (Buffon, 1792, v. 1, p. 10).

Buffon's views on geology changed as he pushed on with his grand writing project, but he did not revise early volumes. Instead, he collected his "additions and corrections" and attached them to a new book-within-a-book, *the Époques de la Nature* (Buffon, 1778), which first appeared as volume 34 of the series, but was hugely successful by itself. It presents a new system, treating the history of the Earth in seven epochs:

1. Since the formation of the Earth and the Planets.
2. Since the matter consolidated and formed the rock inside the globe, as well as the great igneous *(vitrescible)* masses which are at its surface.
3. Since the waters covered our continents.
4. Since the waters retreated and volcanoes began to act.
5. Since elephants and other southern animals inhabited the regions of the north.
6. Since the continents were separated.
7. Since the power of man overcame that of nature.

Buffon had conducted an elaborate series of experiments with spheres of various diameters heated and allowed to cool (Buffon, 1778 Supplément, v. 1, 2). Contrary to his earlier opinion, he had become convinced of the Earth's internal heat, and he thought that he could calculate the ages of planets by extrapolating the observed cooling rates of his spheres. For the Earth he calculates 2936 years for the time it took to "consolidate to the center," some 25,000 years to the time when water first condensed on the surface, and 132,000 years to the present. The whole sequence of events he envisions on the Earth is predicated on its rate of cooling. As a crust formed on the molten globe, irregularities developed and mountain chains were pushed up, as proved by the granitic cores of the highest mountains. Water then covered the entire Earth, as demonstrated by the presence of fossil shells on all but the highest mountains. The sea gradually retreated, depositing fossiliferous limestones and shaping the surface as it went so that the fossils at high elevations are older than those in lower places. Fossil remains of vertebrates are found only in superficial deposits.

Large animals such as elephants, now living only in southern climates, evolved in the mountains of the north as the temperature of those areas first dropped to a level suitable for their propagation; then they moved south as the temperature fell further. That is why elephant remains are found in such places as Siberia and Canada. Man originated in north-central Asia and from there spread all over the world before the continents were separated. There are no elephants in South America because they were unable to cross the high mountains of Central America, but man had no such problem and a race of giants still thrives in South America. The Earth is still cooling, as proved by the Alpine glaciers which he thought were still growing.

Buffon has a way of stating his views with great assurance and a compelling clarity, guaranteed to elicit approval from many readers and violent objections from some. Both reactions helped sales, and the volumes of the *Histoire Naturelle* were printed and reprinted, translated, bowdlerized, imitated, "adapted for ladies," and "abridged for schools" for more than a century and a half. There was even a bibliophile edition with drawings by Picasso published in Paris in 1937.

His intention to take theology out of natural history did not escape the attention of the clergy. The Theological Faculty of the Sorbonne spent two years studying the first three volumes of the *Histoire Naturelle,* and then brought Buffon before a council in 1751, pointing out discrepancies between his text and the letter of Genesis and demanding retraction. Now Buffon was a powerful personage, both politically and financially, and he had the support of Louis XV, but even so he did not quite dare tell the theologians not to interfere. Instead he put on a great show of humility, recanted his errors, and in 1753 published a point-by-point submission in volume 4.

Obviously he failed to mend his ways because 25 years later the Theological Faculty was after him again, for the deviations he had allowed himself in the *Époques de la Nature* (Acta, 1780). Actually, Buffon had had his own doubts about the acceptability of the chronology he had calculated, and he had slipped an apology very much like Descartes' into the preface to the *Époques,* but that was not enough. The theologians objected to his statement that Europe was separated from America when the powerful kingdom of Atlantis foundered about 10,000 years ago (Buffon, 1778, p. 206). They pointed out that only "6,000 to 8,000 years have elapsed since the Creation, as fixed by Moses," and cited two scholarly references to prove it (Acta, 1780, p. 16). Again Buffon "willingly recognized having erred in judgment" (Acta, 1780, p. 19), and again he promised to publish a retraction "at the head of the next volume of my works to be published," but by that time he was older, richer, and even more powerful, and somehow never got around to satisfying that promise. Also by that time, as we shall see, the *encyclopédistes* had openly defied the Church in such matters and had gotten away with it.

THE FIGURE OF THE EARTH

In France, as in England, the early development of what is now geophysics grew out of practical-minded efforts to improve the accuracy of long-range navigation. By the end of the seventeenth century, London instrument makers had learned how to engrave very accurately graduated circles for astronomical instruments, and the new telescopic sights made it possible to take full advantage of that precision. Latitude determinations with an accuracy of a few minutes of arc became commonplace, and now it was necessary to consider the exact shape of the Earth.

Until the development of these accurate circles it was close enough to assume that the Earth was a sphere, just as the ancients had done. Astronomers had been referring their observations to the plumb line and assuming that the line of the vertical passes through the center of the Earth. In France there was also a strange, almost metaphysical fascination with the concept of a "natural" standard of length, to be used in establishing a new rational and universal measuring system. The second, as a unit of time and of arc, seemed perfectly "natural" and the length of a simple pendulum that beat once a second naturally appeared to be the leading candidate for a universal standard of length at about the time when the French Academy of Sciences was getting started.

Very accurate pendulum clocks based on the design of Christiaan Huygens (1625–1695) had been available since about 1657, and Huygens himself had been enticed to move from The Hague to Paris in 1666 to be a major light in the new Academy. Even though his great work on gravity (Huygens, 1673) was still some years off, Huygens already suspected that a standard pendulum would beat slower at the equator than at the poles because of the centrifugal force produced by the rotation of the Earth.

In 1670 the French Academy sent a resourceful astronomer named Jean Richer (1630–1696), first to the coast of what is now Maine and two years later to Cayenne in French Guiana, with the best pendulum clocks available. The Maine trip was not a total success because both clocks had stopped at sea and Paris time had been lost, but it did produce the first accurate latitude determination in America, at the French fort in Penobscot Bay. The Cayenne trip went smoothly: Richer stayed a full year, and made accurate determinations every week for 10 months. He found that a clock adjusted in Paris (latitude 48 degrees 50 minutes) was

2 minutes and 28 seconds a day slower in Cayenne (latitude 4 degrees 56 minutes, as determined by Richer).

The *Principia* of Isaac Newton first appeared in 1687 and contained, among many other geophysically fundamental ideas, a calculation of the figure of the Earth based on its rotation as a body held together only by the force of gravity. It concludes that "the diameter of the Earth at the equator, is to its diameter from pole, as 230 to 229" (Newton, 1729, v. 2, p. 243). Newton was impressed by the "diligence and care" of Richer's observations and cites them as primary experimental proof of the theoretically calculated flattening (*Principia*, book 3, Proposition 20).

In the meantime, Gian Domenico Cassini (1625–1712), another big name brought to Paris by the new Academy (this time from Bologna), began to organize the great project of extending the meridian of the Paris observatory across all France by ground survey and making astronomical observations of latitude at points along it. It took a long time, but by 1718 an accurate triangulation network extended from Collioure in the south to Dunkerque in the north. The results of the astronomical observations led Gian Domenico's son and successor as Director of the Paris Observatory, Jacques Cassini (1677–1756), to announce "definitively" that the length of a degree of arc on the surface of the Earth decreased toward the poles (Cassini, 1722). The Earth, he asserted, was not flattened at the poles as Newton had calculated, but was pointed (like an American football), and somehow that was construed as a great victory for Cartesian physics over the newfangled Newtonism. It was a matter of observational inaccuracies, but as a subject for popular scientific debate it fitted nicely with the widely fashionable commentary that surrounded Newton's *Principia*.

The contemporary popular criticism of the *Principia* was rarely illuminating. Some writers praised the work and others condemned it, but except for a small group of mathematicians and astronomers, few commentators were able to show that they had understood it. The concept of attractive force as a fundamental property of matter was admittedly difficult to grasp. Newton was a great master of precise language, but his manner of presenting mathematical relationships may have been obscure even to readers with some mathematical training. His Latin must have been a joy to read, compared with the quasi-Latin that was current at the time, and his grammar was perfect, but the power of his new physics was not obvious.

Among the few who had understood the *Principia* was a trio of lively Frenchmen, Pierre Louis Moreau de Maupertuis (1698–1759), Pierre Bouguer (1698–1758), and Alexis-Claude Clairaut (1713–1765). All had started out as mathematical prodigies, all had become avid Newtonians, and all had great trouble explaining to their elders why the *Principia* was so important.

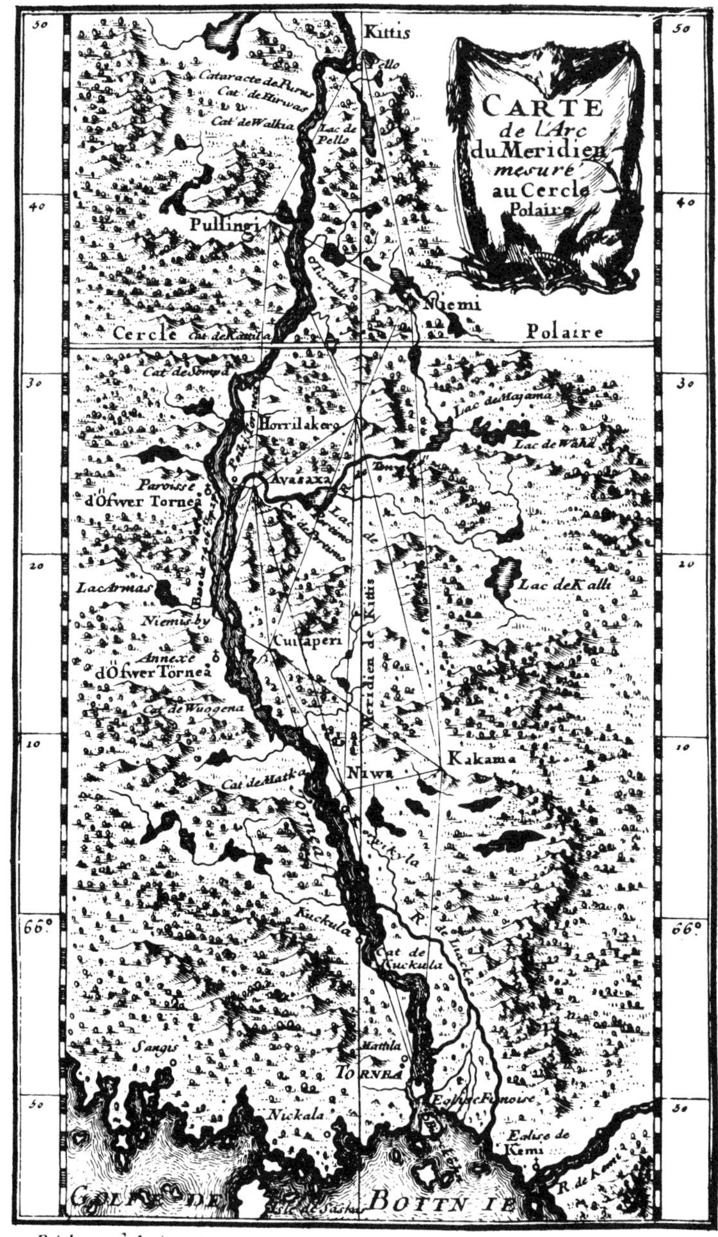

FIGURE 5.2. Triangulation from Tornio to Kittis Peak, to measure the length of a degree of latitude at the Polar Circle and thus determine the flattening of the Earth. (Maupertuis, 1738)

In 1733 Maupertuis had detailed again how the Earth's flattening could be verified by long-range triangulation along the line of a meridian, coupled with astronomical determinations of latitude at both ends of the network. He had been to London five years earlier and knew that instruments of appropriate accuracy were available there. The Academy agreed and in 1735 sent a team headed by Charles-Marie de La Condamine (1701–1774) to Ecuador (then part of Peru under Spanish rule) to make those measurements. Pierre Bouguer went on that expedition and had a terrible time. The terrain was rough, overland travel was difficult, the Spanish authorities were suspicious and uncooperative, and the senior scientists of the expedition spent much of their time fighting each other. It took nine years to finish the triangulation from Quito to Cuenca and to make the astronomical observations at both places. Bouguer did not get home until 1744, and by that time the question of the Earth's flattening had been resolved by the second expedition which had gone to the Arctic Circle in 1736.

That second expedition lasted only 14 months and returned to Paris in August 1737. Maupertuis was in charge, young Clairaut came along, and Charles-Étienne-Louis Camus (1699–1768), Louis-Guillaume Le Monnier (1717–1799), and Réginald Outhier (1694–1774) were the rest of the French team. Anders Celsius (1701–1744), professor of astronomy at Uppsala (who later gained immortality with a paper on thermometer scales), happened to be in Paris in 1735, offered his assistance, and became a prime mover of the expedition. They picked the valley of the Torne River (now the boundary between Sweden and Finland), running almost straight north across the Arctic Circle, and surrounded by mountains that were high enough to permit long-range triangulation, with some shots longer than 12 miles.

The provincial governor was very friendly, the Finnish boatmen showed great skill on the river, and a detachment of Swedish soldiers did most of the heavy work in the dense northern forest. They triangulated 60 miles up the river, lugged their 9-foot sector and many other instruments to the observatory they built on Kittis Peak at the northern end of the traverse, determined the latitude down to a few seconds, then hauled it back again to Tornio (near the mouth of the river) to repeat the procedure at the southern end. They laid out the baseline for their triangulation in early winter on the ice of the river, 14,812 yards with a precision of a few inches. Then they spent the long dark winter calculating and recalculating their results, which showed that the Earth was flattened a little less than Mr. Newton had predicted, but close enough.

Maupertuis presented the results in his *Figure de la Terre* (Maupertuis, 1738), and there he also gives an exciting account of the expedition's adventures. Clairaut mulled over the problem some more and then came out with his *Théorie de la Figure de la Terre* in 1743, where he gives the theorem that now bears his name, relating the pull of gravity to the latitude. Bouguer could only confirm all that when his *Figure de la Terre* finally appeared (Bouguer, 1749), but he managed to add one new observation, made in the high Andes: The lessening of the pull of gravity with altitude is partly compensated by the gravitational attraction of the intervening rock. This is what we now call the Bouguer correction.

THE *ENCYCLOPÉDIE*

One of the fixtures of the *Jardin du Roi* was (and still is) a lecture hall where professors gave a variety of courses and the public was admitted free. In 1742 Buffon brought in as professor of chemistry a pharmacist who had been making quite a reputation for himself with his private lectures, Guillaume-François Rouelle (1703–1770), a friend and probably a student of the Jussieu brothers. He was a practical man, his lectures were substantive, dynamic, and interesting, and the best minds of Paris had been crowding his house to hear them. Now he had an even wider audience, from all walks of life (Rappaport, 1960). Young Hutton was in Paris from 1747 to 1749, probably heard Rouelle lecture, and one could speculate that his interest in geology was kindled there.

Rouelle published neither a text nor a syllabus, but the substance of the lectures he gave about 1754–1757 is perserved in a good set of notes taken by his friend Denis Diderot (1713–1784), the chief *encyclopédiste*. From this manuscript it appears that Rouelle's chemistry was scientifically as advanced as any that could have been found in Paris at the time, and that it incorporated a lot of mineralogy and geology.

In a wider sense it is difficult to gauge the full dimension of Rouelle's importance to French geology. Distinguished people (Diderot, Malesherbes, Rousseau, and Turgot, among others) are known to have come to hear him, and some very important scientists (notably de Saussure, Desmarest, and Lavoisier) took the more detailed courses he gave in his shop for paying students, but his influence may have gone far beyond that. Several *encyclopédistes* acknowledge his help in their research, and he seems to have been close to d'Holbach and his translations of Wallerius' *Mineralogia*, Johann Lehmann's treatises on mining technology and rock strata, and several important German works in chemistry. His outlook on local stratigraphy was still horizontalistic, but on a larger scale he followed Lehmann's idea of a basement overlain by sequential stratigraphy. He saw the importance of chemistry in mineralogy as put forth mainly by Wallerius in Sweden. We have no lists of attendance in his courses, but one could imagine that many future greats had come to hear him. Rouelle's lectures may have been the main inspiration for most of French (and some German and British) geology of the second half of the eighteenth century.

The French *Encyclopédie* became the embodiment of the Age of Enlightenment. It began as a commercial translation of Chambers' *Cyclopedia*, which had enjoyed great success in England since 1728, but took on new directions when the publisher engaged Diderot and d'Alembert as editors in 1747. By that time Diderot was recognized as a writer among the Paris *philosophes*, but had not yet made much of an impact on the national scene. Jean le Rond d'Alembert (1717–1783), who was to be the science editor of the new venture, was already well known as a mathematical physicist, mainly through his *Traité de Dynamique* in 1743, an important work in Newtonian mechanics. Imbued with the spirit of free inquiry, social justice, and the science of natural causes, the editors made it their business to assemble far more than a mere summary of contemporary knowledge. They advanced progressive views on all aspects of existence and frequently took political positions that were anathema to the Church and the State alike.

Diderot was able to enlist the finest minds in France as his collaborators, and he kept after them until they delivered. The first sumptuous folio volume appeared in 1751 and six others followed in good order, but the liberal and irreligious tone of the work was too extreme for the conservative bishops and nobles of the *ancien régime*. Actually the Church was far from unanimous on this issue, and more than a few abbés were happy to write articles for Diderot, but the conservative opposition found it easy to occupy the limelight and create a disturbance.

In 1759 the project's royal *privilège* was ceremoniously withdrawn, in Italy the book was placed on the Index, and Pope Clement XIII ordered all faithful to bring in their copies and have them officially burned (Lough, 1968). That could have been the end except that the *Directeur de la Librairie*, the czar of French publishing, at that time was Chrétien-Guillaume de Lamoignon de Malesherbes (1721–1794), whose father had been Chancellor of France. Malesherbes was an accomplished naturalist, had studied with Bernard de Jussieu (Antoine's partner and younger brother) and with Rouelle, and now applied his considerable administrative talents (and his great influence) to keeping the *Encyclopédie* afloat. Discouraged, d'Alembert had resigned, but Diderot continued behind closed doors with Malesherbes' protection, and the printing went on in Paris until the 10 remaining volumes of text were finished in 1765. Then they were released all at once, over a false Neuchâtel imprint, and the deed was done. By that time it had long been obvious that the *Encyclopédie* was a huge success. The high quality of the articles, the unfettered approach, and the enlightened outlook combined to make it "the biggest best-seller of the century" (Darnton, 1979).

While this was going on, a new *privilège* was obtained for a book entitled *Recueil de Mille Planches* (Collection of 1000 Plates), which happened to be the illustrations for the *Encyclopédie*, and they continued to appear legally from 1762 to 1772. In the end there was not 1000 of them but almost three times that many, published in 11 volumes, their lavish copper engravings depicting every aspect of the work of man and nature. Diderot was particularly interested in the trades and in their technology, and the drawings he put together now comprise our best record of how things were done in the eighteenth century.

Illustrations of geological subjects appear in volume 6 of the *Recueil*, published in 1768. They were assembled and seen through the press by Paul Henri Thiry, Baron d'Holbach (1723–1789), a modest German nobleman who lived most of his life in Paris. He was well read in the contemporary German literature in chemistry, mineralogy, and geology, and as a translator had the critical ability to recognize what was new and important, but his accomplishment as an art editor did not measure up to his other talents. Most of the drawings in his volume were compiled from other books going as far back as Agricola, and some of the copying was inaccurate. His plates of fossils are poorly drawn and some of his mining diagrams are not as clear as the original woodcuts in *De re Metallica*, but the volume is important because it contains a great geological breakthrough: Desmarest's plates of prismatic basalts and his discovery that all basalt is volcanic.

D'Holbach was perhaps the hardest working and the most militantly irreligious of Diderot's science writers. His house was a central meeting place for the *philosophes* and a hotbed of radical views of science, religion, and politics. They had all read Lucretius and gone on from there. In such a dynamic interchange of opinions and ideas it is now difficult to decide who influenced whom in what way, but it is clear that the interactions raised the level of all. D'Holbach, for example, learned from Rouelle and also taught him. *Telliamed* was their bible, and when the new edition of Palissy appeared, they espoused it too. They quoted the great Buffon right and left, but made fun of his popularizations. Unlike him, they had the courage to stand up to the attacks from the "first estate," the high clergy.

Over the years d'Holbach wrote close to 1000 articles for the *Encyclopédie* (Lough, 1968, chapter 3) on chemistry, mineralogy, geology, mining, and metallurgy, as well as geography, philosophy, religion, and what today would be called anthropology. He was more interested in transmitting ideas than in taking credit for them, and most of his writing does not have his name on it. His famous *Système de la Nature* (d'Holbach, 1770), appeared in Amsterdam, over a false London imprint, and under the pseudonym M. Mirabaud (which was the name of the *secrétaire perpétuel* of the Academy, the living symbol of the French Establishment). In it d'Holbach builds a materialistic system controlled only by natural laws, rejects all interpretations based on faith, and postulates an ecological transformation of species, including man, on a dynamically self-regenerating Earth. It was an audacious conception, hazy on detail, but presenting a sharp separation of science from religion—a thought that did not find publication in England until 60 years later.

BASALT

While Réaumur and Buffon still held center stage and the Holbachians were furiously scribbling for Diderot, a friend of theirs was beginning a long career that was to have great influence on French geology. He was Jean-Étienne Guettard (1715–1786), a botanist and naturalist by training and assistant to Réaumur. In 1747 he became physician to the Duke of Orléans and, following the Duke's interests, turned to geology. When the Duke died five years later, Guettard received a pension and rooms at the Palais Royal, which made him independent and free to continue his geological research.

In 1751, Guettard and Malesherbes made a journey to the Auvergne (in central France). The original purpose of the trip was to study local industries, and that led them to inspect some elegant and elaborately carved old fountains in Moulins, on the road to Vichy. The fountains were made of a black stone that appeared to be very durable but was obviously easy to work—a kind of stone they had never seen before. They were told the stone had come from Volvic (north of Clermont) and they went to the quarry. Neither of them had ever seen a volcano, except in books, but they soon realized that this stone was volcanic and that it was being quarried from a large lava flow (now described as consisting of pumiceous trachyandesite). All around them they noticed craters of volcanoes, long peaceful and covered with woods. They had discovered a region of extinct volcanoes right in the middle of France (Ellenberger, 1978)! That was the first of the great studies which established the Auvergne as a classic geological area, destined to become the "graveyard of neptunism." It is amusing to contemplate that Abraham Gottlob Werner, the apostle of neptunism, was still a baby when Guettard and Malesherbes first recognized the volcanoes of the Auvergne.

In spite of these keen observations, Guettard continued to believe that the heat of volcanoes comes from combustion not far below the surface, thus following the early view of Buffon, whom he had severely criticized for reaching unwarranted conclusions. For more than 20 years after the trip to the Auvergne he continued to hold with the neptunist opinion that prismatic basalt precipitated from sea water and that the polygonal cracks were due to desiccation, such as mud cracks.

FIGURE 5.3. One of the ancient fountains in Moulins, made of the "stone of Volvic," a material Guettard had never seen before.

It was a matter of misplaced evidence. Crystallization from solutions was obvious to eighteenth-century chemists, but every time they melted granite or basalt, the melt cooled to a glass—a substance quite different from the original rock. The crystals in granite were obviously insoluble in water yet they appeared similar to crystals of common salts, grown in the laboratory, and it seemed reasonable to assume that they also must have crystallized from a solution in water. The traditional emphasis on water as an agent in the process of Creation supported the assumption.

A GEOLOGIC MAP

Collecting information on nature had been a valid scientific pursuit for some time. Lhwyd and Woodward had obtained much useful geological data from the questionnaires they had circulated, mostly to country parsons and gentlemen in Britain. The idea of plotting such information on a map had occurred to Lister (1684), but as far as is known, he never took it beyond a proposal. The French Academy of Sciences had been collecting environmental information for decades, and the *Encyclopédie* was a splendid product of such efforts, but Guettard was the first to attempt presenting collected geological data on a map.

He became aware of the topographic regularity with which geological formations followed each other in the Paris basin and realized that a graphic representation of the way the rocks were distributed in the region could be helpful in the search for useful minerals and building materials. He used various symbols, mostly borrowed or adapted from the chemists, to show mineral localities, rock types, and fossils, and he sketched in broad bands to indicate different country rocks, without any attempt to show stratigraphic superposition. His first *Carte Minéralogique* covered England and France and was presented to the Academy in 1746 (Guettard, 1751). It was followed by similar maps of the Middle East (Guettard, 1755), Switzerland, eastern North America (Guettard, 1752*a*, 1752*b*), and Poland (Guettard, 1764)—all compiled almost entirely from other people's observations. Guettard himself never set foot in the Middle East or in America.

In 1766, Guettard and his young friend Antoine-Laurent Lavoisier (1743–1794), the great chemist in later years, entered into a project sponsored by the Minister of Mining to make a geological survey of all of France in 214 quadrangle maps. It was an overwhelming proposition. There was no agreement on what the ultimate purposes of a geologic map should be or what it might usefully show. The government's intentions were progressive, but support was meager and halting, and the mapping turned out to be much more work than anyone had expected.

Guettard and Lavoisier at times disagreed on how maps should be made. Guettard was fundamentally opposed to theorizing and was afraid that if he made stratigraphic interpolations across the grass-covered landscape, he would be erecting only more theory. Lavoisier had learned how to recognize a number of formations in the field, was beginning to understand their stratigraphic sequence, grasped the importance of that information, and wanted to show it on the maps. Guettard could see his point and allowed him to draw stratigraphic columns in the right margin of the maps, but nothing on the maps themselves.

After 11 years of work they had finished 16 quadrangle sheets and had about a dozen more in various stages of completion. Lavoisier was becoming involved in other things, both science and business, and seems to have drifted away from the project. In 1777 the survey was taken over by the Inspector General of Mines, Antoine-Grimoald Monnet (1734–1817), a student of Rouelle, and he published an atlas of 31 maps in 1780. A later version with 45 maps appeared without date and is known today in only one copy. (For an excellent discussion see Rappaport, 1969.)

It was the first national geological survey, but for a half-century afterward it appeared that it might be the last. The techniques of geologic mapping developed rapidly in the late eighteenth century in France (see, e.g., Desmarest, 1804, 1806; Soulavie, 1780–1784; Cuvier and Brongniart, 1808–1811), but the usefulness of geologic maps was obvious only to a few specialists and then was directed mainly toward specific geological problems. To the public and to the governments, organized geological surveys did not appear to be particularly worthwhile.

Guettard published about 100 scientific papers, many of them long and dull, but he included important insights. His seminal ideas are often buried in meticulously recorded but seemingly endless detail and his great fear of generalization and theorizing occasionally kept him from fully interpreting his own evidence. The papers were not often cited, but it seems likely that his personal influence on the naturalists in the *Jardin du Roi* was greater than the credit they deigned to give him in their publications. His crucial position in French geology was realized only much later (Geikie, 1905).

One of the first to follow Guettard into the Auvergne, in 1763, was Nicholas Desmarest (1725–1815), then Inspector of Manufactures of the Generality of Limoges (and later of all France), an important cog in the centralized industrial system and an amateur geologist. Having read his *Telliamed* and studied with Rouelle, he naturally leaned toward neptunism at first, but once he had seen the volcanoes of the Auvergne he greatly modified his attitude. He continued to accept the importance of water as a formative agent in the Earth's history, but if we agree that the Auvergne is the burying ground of neptunism, then Desmarest becomes the unwitting undertaker (however, it took a long time for the geological community to realize that the funeral had taken place).

Desmarest's discovery was direct and simple: Near the Mont-Dore (southwest of Clermont, then called Mont d'Or) he found prismatic basalt forming part of an obviously volcanic lava flow. It was clear that the hexagonal cracks in the stone were the result of cooling, not desiccation. He presented his stunning evidence to the Academy in 1765 and was off to Italy on an official mission. His observations excited the Holbachians, and they published them in an unusual place: a page of text written in the third person, spliced into the list of plates for d'Holbach's volume 6 of the *Recueil*, and accompanied by two handsome original plates showing prismatic basalt in the Auvergne. The result was noticed, and it earned Desmarest's election to the Academy even before it was formally published (Desmarest 1774, 1777), but not many were ready to believe it. Guettard himself did not see the volcanic origin of prismatic basalt until about 1775 (Guettard, 1779), and the German and Scottish branches of the neptunist family did not give up until a half-century later. Desmarest survived the Terror and lived through most of that time but declined to take part in the basalt controversy. [Geikie (1905) gives a fine appraisal of Desmarest's geology.]

In 1764 Desmarest began the detailed mapping of the volcanic area from Volvic to the Mont-Dore, supervised the project for several years, and continued to refine the map for the rest of his life. That was probably the first time that a fairly small area was mapped on a large scale to illuminate a geological principle which applies generally (Desmarest, 1806). Again, the principle was simple: Lavas from volcanoes flowed downhill and followed valleys that had been cut by rivers. Those same rivers then began cutting new valleys into the lava flows that had blocked them, and they are still doing it. The topography had been shaped by running water in the past, just as it is today. Old basalt flows now may be capping flattop hills, but originally they were in valleys or on plains, and all the hills formerly around them (including the volcanoes from which they had come) have now been removed by erosion. The word *uniformitarianism* had not yet been coined, and Desmarest saw no reason for making an issue of an idea that seemed so obvious to him.

The studies of volcanic phenomena were extended beyond the Auvergne by Barthélemy Faujas de Saint-Fond (1741–1819) into the Vivarais (Ardèche) (Faujas, 1778) and into Scotland (Faujas, 1797). He had been a lawyer, but in 1778 became assistant and later professor of geology at the *Jardin du Roi*. A vivid and accurate description of Italian volcanoes was given by Déodat de Gratet de Dolomieu (1750–1801), a Knight of Malta, contentious adventurer, aristocratic partisan of the Revolution, and one of the best geological minds of the time. He realized that the heat of volcanoes could not come from any combustion and searched vainly for alternate explanations, but like Desmarest, he could not "go back" to the flaming Earth of Descartes (Dolomieu, 1783, 1788). Dolomieu also did important work in mineralogy; a calcareous mineral he had described from the Tyrolean Alps (Dolomieu, 1791), and later the Dolomite Alps themselves, were named after him.

Late in life Dolomieu was appointed professor of mineralogy at the *Jardin du Roi* and in 1802 was succeeded in that post by his friend, the Abbé René-Just Haüy (1743–1822). Building on the ideas of his Paris colleague Jean-Baptiste Louis Rome Delisle (1736–1790) and the Swedish chemist-mineralogist Torbern Olof Bergman (1735–1784), Haüy developed a system of mineralogy based on crystal symmetry and chemical composition, which clearly established the fundamental importance of the internal structure of crystals in all mineral classification (Haüy 1784). He spent his life refining the systematics of mineralogy and lived to see his scheme widely adopted. In that sense he was perhaps the most permanently influential voice to emanate from what had become under the Republic the *Jardin des Plantes*.

THE TWO QUESTIONS

The volcanoes of south-central France had been discovered and the igneous origin of basalt demonstrated, but the central importance of heat as a geological factor was not established in France. Even to the volcanists, water remained the fundamental agent that deposits rocks and shapes their surface. Volcanism was fascinating, but wanting an acceptable explanation for the origin of its heat, it had to be regarded as exceptional and therefore unimportant on a global scale. The nature and origin of granite remained a total mystery.

The horizontalist view that different strata had been deposited in different parts of the same sea at the same time had been put in serious question by Lavoisier in his stratigraphic columns in the margins of some of Guettard's maps, but that applied mostly to the Paris basin. In southern France, where sedimentary rocks are mostly limestones that appear to slope gently toward the Mediterranean, it was easy to agree with de Maillet and think of them as the residual deposits of a diminishing sea.

So it was that the two basic problems: (1) the relative ages of sedimentary formations, and (2) the role of heat in large-scale geological processes, remained unsolved. An imaginative country abbé in southern France tackled them both and came close to resolving them. He was Jean-Louis Giraud Soulavie (1752–1813), one of the bright and liberal thinkers in the Church with early sympathies for the Revolution, who moved from the vicarage of Antraigues (Ardèche) to the salons of Paris, left the Church, married, took an active part in the Revolution, and in time grew to detest it. He met Benjamin Franklin in Paris, they became friends, and in 1786 he was elected a member of the American Philosophical Society.

The geological phase of his career was brief, about 1775 to 1780, and he was still in his twenties when he presented his results in an 8-volume work on the natural history of southern France (Soulavie, 1780–1784). He came no closer than his colleagues and predecessors to understanding the origin of volcanic heat, but his field mapping convinced him that volcanoes are more than local curiosities. His maps still use the kind of symbols Guettard had introduced, but they also show colored areas with boundaries to distinguish the "granitic terrain" from the limestones *(terrain calcaire)* and the regions affected by volcanism *(terrains volcanises)*. He makes no age distinctions on the maps, but in the text he repeatedly explains the concept of superposition, and carefully analyzes its significance in determining the relative chronology of geological events, whether sedimentary or volcanic. He recognizes the changeability of fossils in geologic time and their consequent utility in correlation. His stratigraphy is crude and replete with local mistakes, but he was the first to build on the ideas of Arduino and Lehmann and develop the relative time relationship of formations by the fossils found in them.

His own presentations of his chronology are characteristically inconsistent, but they boil down to a fivefold division. The oldest unit, presumably resting on "granite," he calls *Premier âge* (or "primordial" or *marbre*), and it consists of dense limestones and "marbles" bearing fossils that have no equivalents in the present seas (like ammonites and belemnites). The *Second âge* (or *secondaire*) is also a limestone and it also contains the remains of animals now extinct, but mixed in with them are shells similar to species now living. The rock itself is softer and made up of thinner beds than the underlying "primordial" formation. The *Troisième âge* (also called *pierre blanche*) is "the exclusive realm of (petrified) shells now living in our seas." The *Quatrième âge* is the "realm of fishes & of plants known today" and the *Cinquième âge* is characterized by "petrified trees, conglomerates, bones of fossil animals, &c." (Soulavie, 1780, v. 1, p. 317).

Soulavie observed that some species must have changed in various ways over the course of time and others disappeared entirely, which made his picture of Creation differ substantially from the official version. He also understood that his fivefold chronology required much more time than biblical chronology would allow, and here and there in his text he injected suitable apologies and assertions of "perfect agreement" of his geological conclusions with Genesis, but his irreligious attitude was transparently obvious. It led to a long disagreement with his ecclesiastic critics and with the Academy (Aufrère, 1952), a struggle he did not quite win and which contributed to his decision to leave the Church.

The geological structure of southern France is far too complicated to allow the sorts of generalizations he was trying to make on such limited experience. He understood the methods of historical geology and he pointed the way to stratigraphic correlation by fossils, but the written expression of his conclusions is disorganized, often unclear, and repeatedly inconsistent. He did not realize that the oldest rocks in his system were geologically still quite young, and ultimately he failed to document a coherent scheme. Thus he left it to the British to integrate these ideas and to bring them to fruition.

CHAPTER SIX

The Detour Via Freiberg

THE GERMAN MINING ACADEMIES

By the middle of the eighteenth century, the German mineral industry had advanced a long way beyond the technology described by Agricola (1556), and his elegant textbook was finally becoming obsolete. Politically, Germany was fragmented into a patchwork of more or less local governments with a diversity of often conflicting foreign affiliations, but the mining community continued to hold together across all the political boundaries and thus continued to provide not only the metals essential for the governments' coinage but also a measure of national cohesion. Business was better than ever, but new deposits were becoming increasingly difficult to find and the average grades of the ores being worked were steadily decreasing.

The spirit of rationalism demanded a technical approach to these problems. The leaders of the mining industry understood the need for technical training of the new generations of mine surveyors and mineral engineers, but German universities, like their sister institutions in England, paid little attention to the new trend. At the same time, scientific societies were springing up everywhere and reporting obvious technical advances in their widely diverse publications. Occasional scientific lectures in the universities themselves confirmed the growing interest in science and technology, but courses reflecting that interest rarely appeared in the curricula.

Unable to penetrate the universities, the German mining community set out to organize its own schools to teach engineering, assaying, and the mineral sciences. The first mining academy was established in Prague in 1762 (Peithner, 1780), and others soon followed at Freiberg in Saxony in 1765, in Berlin in 1770, and in Clausthal-Zellerfeld in 1775. After 1770, the Prague academy moved to Schemnitz (now Banska Stiavnica in Slovakia) and consolidated with the forestry school that

FIGURE 6.1. Mining Academy students, in uniform, surveying a line. The city of Prague is in the background, and a monument to the Empress Maria Theresa is at the right. (Peithner, 1780)

had been there since 1660 (Kettner, 1967); the other three still operate where they began.

The sense of new directions was still very fresh in Freiberg when Abraham Gottlob Werner (1750–1817) was called there at the age of 25 to teach mineral science. He had been there as an undergraduate and now had just finished three years of miscellaneous studies in Leipzig which could have qualified him as a mining lawyer, but he was more interested in minerals. He had come from a line of Saxon ironmasters and had been involved with mining and mineralogy since childhood. While in school in Leipzig he had written a book on mineral identification (Werner, 1774), which is what got him the job.

The book is basically a mineralogist's field manual and presents a collection of descriptions and definitions of the external characteristics of rock and mineral specimens intended to be helpful in their identification by sight. Color, luster, fracture, and gross morphology are discussed in great detail, but physical properties are defined only in a qualitative way, with little suggestion that they could be measured.

Even specific gravity is stated in comparative terms ("floating, light, not particularly heavy, heavy, extremely heavy") and hardness has only four steps ("hard, half-hard, soft, very soft"). Limestone and basalt are both listed as examples of "half-hard."

Werner was aware that density could be measured accurately, but he preferred to leave that to the physicists. He felt that the mineralogist would do better just hefting the specimen in his hand because the physical properties were unreliable anyway. The *Mischung* (literally "mixture," meaning the chemical composition) of minerals is essential in their classification, he admits, but much too difficult for the mineralogist to determine. Chemistry is "a science that is not yet fully developed," he writes (Werner, 1774, p. 40). He classifies the gross aspects of crystals, but the significance of their forms eludes him. He was not aware of the recent work of Romé De L'Isle (1772).

Now we may find it surprising that such a homely little book would have had so much influence, but in its day it was welcomed as a great step out of the muddle of inconsistent classifications and conflicting nomenclatures that had mired the study of rocks and minerals since before Theophrastus. Werner's approach was orderly and logical. In his mineral names he had followed the usage of German miners, and they were a large constituency. They appreciated Werner's elementary presentation and his disdain for chemistry and physics. The book was written in their language.

At the Freiberg Academy they regarded Werner as a hometown boy who had succeeded in the big city and who would do great things for the school. He did not disappoint them. He organized his courses with boundless energy, fascinated the students, and soon was the star of the *Bergakademie*. One may read that his fame spread all over the world and that Freiberg became a mecca for geology, but that is only partly true. The school was doing nicely, Werner's courses were well attended, and many of his students were coming from abroad, but the Academy was a small, narrowly based, provincial operation. In spite of the glowing reports of Werner's pupils, Freiberg never attained the prestige or the enrollment of a major university. When the mineralogist Friedrich Mohs (1773–1839), Werner's student, his successor at the Academy, and the author of the decimal hardness scale, was offered a post in Vienna in 1826, he took it as a great promotion and promptly left Freiberg.

Werner did not like to write, either privately or for publication, and on the rare occasions when he finally forced himself to put together a manuscript, he hated to let it go out. Much of what we know about his geological ideas thus comes from the reports published by others (mainly Jameson, 1804–1808), generally based on his lectures, and a few of his own manuscripts or notes published without his permission or after his death. Besides a few brief and mostly polemical articles and a catalog of a large mineral collection (Werner, 1791*a*), Werner himself published only two major works during his Freiberg career. The first (Werner, 1786) was a classification of rock types according to his system of sequential precipitation and deposition from the waters of the universal ocean, and it is there we find his first assertion in print that basalt was deposited from water. The second (Werner, 1791*b*) was a theory of veins which are, in his scheme, the often ore-bearing formations that cut across the bedding of layered rocks. The veins, in his view, are nothing but open tensional fractures in the rocks, produced mainly by the shrinkage that follows deposition, and then filled from above by material precipitating from the waters.

WERNER'S GEOGNOSY

Werner's ideas changed little during the four decades of his teaching. He maintained that all *Primitive* rocks crystallized from the waters of a primordial ocean. Granite came first, followed by gneisses and slates, granular limestones, and a variety of other crystalline rocks, including "slaty glance-coal" (Jameson, 1805, p. 83, from the German *Glanzkohle*, meaning anthracite)—all deposited on the uneven bottom of the sea. The Primitive ocean was devoid of life, and Primitive formations contain no fossils. They are overlain by a series of *Transition* formations, "first established as a distinct class by the acuteness of Werner" (Jameson, 1808, p. 145). Actually, he was forced to add that category after fossils were found in some rocks he had previously classified as Primitive, and he defined it as formed of rocks that are generally hard but less granular than Primitive strata. The oldest Transition rocks are limestones, usually massive, and they rest directly on Primitive "clay slate" and may contain fossils of the kinds that "gradually disappear in newer formations" (Jameson, 1808, p. 146). Next came the *Floetz* formations, usually found "at the foot of primitive mountains" and consisting of sandstones, limestones, gypsum, salt, "trap," and coal, in a definite, universal order. (Floetz is a German mining term for flat-lying rocks and was introduced into the stratigraphic literature by Lehmann.) "The floetz-rocks contain a great variety and number of petrifactions of animals and vegetables; and these . . . increase in variety and quantity the newer the formation" (Jameson, 1808, p. 154).

The youngest rocks belong to the *Alluvial* period and are composed "of those rocky substances that are formed from previously existing rocks" (Jameson, 1808, p. 206). They fall into two distinct categories: "1st. Those formed in mountainous countries, filling valleys. 2nd. Those formed in low countries, where they sometimes constitute hilly land" (Jameson, 1808, p. 207). "True" *Volcanic* rocks are classified in a fifth category, and they consist of volcanic "ejections" (including some limestone), lava, and "the matter of muddy eruptions." In Werner's scheme of things they are unimportant on a global scale.

To reconcile his system with field observations, Werner had to postulate a complicated history for the universal ocean, making it calm in the Primitive period and stormy in the Floetz, for example, and making it first deep, then shallow, and then deep again, to account for the Floetz rocks found at high elevations. In the end the system was made flexible enough to accommodate any field evidence by assuming regional variations in the sea.

Werner's system was based on a distillate of the views that had been current in Germany when he was a student, collected from Moro, Arduino (via Ferber), Johann Gottlob Lehmann (1719–1767), and Georg Christian Füchsel (1722–1773), among others. Werner called it *geognosy*, a word derived from the Greek *gnosis*, meaning knowledge, and previously used by the German chemist and mineralogist Johann Heinrich Pott (1692–1777) (Pott, 1746). Werner was convinced that his own teaching was based on real knowledge, as opposed to geology, which merely discusses these things (*logos* meaning word or discussion). He took credit for discovering that geological formations are universal and extend around the whole globe onionlike, with few exceptions. Nowhere in print did he mention God, Creation, or Moses, and it appears that his personal religious views were never directly introduced in his lectures, but the system of his geognosy was easily compatible with Genesis, which was a major reason for its widespread success.

Above all else, the personal charm of Werner must have been magic. His lectures could only have been superb, judging by the extravagant tone in which his former students praised them, and the devotion he commanded was astonishing. It is revealing to read the agonies of two of his best students, Leopold von Buch (1774–1853) and Jean-François d'Aubuisson de Voisins (1769–1843), as one after the other visited the volcanoes and the basalt flows of the Auvergne and were "amazed and perplexed" to see so many things so obviously different from what their beloved teacher had led them to expect [(Buch, 1802) in Buch, 1867, p. 518].

The spirit of searching inquiry that pervaded universities and research centers in France and in Britain was less well developed in Freiberg, where the professor spoke and the students absorbed. It was unthinkable to be at variance with the master, much less to contradict him. That was how Werner's theories managed to survive, as long as he was there to expound them. His system was usable and appeared reasonably predictive for a time, but his dogmatic insistence on what were fundamentally false criteria, together with his doctrinaire approach to field evidence, certainly clouded the geognostic picture. At the same time, his brilliant lecture presentations and his infectious scientific enthusiasm called attention to geological questions and made them interesting. He himself produced little that was original, but many students turned to geology because he had attracted them, and made them see his ideas as gospel. It is like detours that lead travelers to see things they otherwise might not have seen, and then, after some temporary confusion, let them proceed to their original destination, perhaps better informed than they otherwise would have been.

CHAPTER SEVEN

The Quiet in Britain

Georgian Britain enjoyed stability at home and expansion abroad. Competition from France was growing, but Britain held her own. Occasional difficulties, including the armed uprising in the American colonies, were only small downward ripples on the steadily rising curve of British influence in the world. The terror of the French Revolution scared the British and the rise of Napoleon caused further discomfort, but in time the upstart was dealt with at Waterloo and Britannia went back to ruling the waves.

For British geology, the eighteenth century had been a protracted period of quiescence. The geological work of the Royal Society had declined under Newton's overwhelming presidency, and the great new ideas of Robert Hooke failed to find even modest acceptance. The Philosophical Transactions of the Royal Society reported Continental geologic intelligence from time to time, and important books from Europe inevitably appeared in English translation, but not always promptly. *Telliamed*, for example, was printed in English only two years after it had first appeared in French, but Buffon's *Natural History*, which made its appearance in French in 1749, did not reach the English market until some 30 years later. The language barrier probably would not have made much difference. Many educated Englishmen read French, but it was more a matter of attitude: Britain knew what was going on in Italy, France, and Germany, but did not choose to find it exciting.

It was a time of geological orthodoxy. Doctors, divines, and educated gentlemen traveled through the countryside, collected minerals and fossils, and observed geological phenomena, often rather well. They sent their observations to the *Philosophical Transactions*, but made little attempt to interpret what they found. They organized their cabinets as best they could and mildly debated whether their "figured stones" were *sui generis* or perhaps the product of the Universal Deluge. In their discussions they cited Burnet and Woodward, together with Theophrastus,

Pliny, and Aldrovandi, but paid little attention to the paleontological work of Ray and Lhwyd, and acted as if Hooke had never existed. In British geology there was nothing comparable to *Telliamed* or the *philosophes* or the *Encyclopédie*. The thought of doubting Scripture, or of going to cliffs and quarries for evidence that might conflict with it, was unthinkable.

THE EARTHQUAKE

The disastrous Lisbon earthquake of November 1, 1755, was felt over most of Europe and sparked much interest in geological phenomena. It was the subject of a flood of sermons, pamphlets, and books reciting the destruction of the city, describing the horrors of the event, and drawing a variety of conclusions, almost all of them theological. A remarkable exception to this literature is a long article in the *Philosophical Transactions* (Michell, 1760), also published as a short book by John Michell (?1724–1793), who is now remembered mainly for his work in stellar astronomy, but who was also an unusually perceptive geologist and served as Woodwardian professor for two years (1762 to 1764).

FIGURE 7.1. The Lisbon earthquake. (von Haller, 1756)

After the Lisbon earthquake Michell analyzed all the information on earthquakes he could find and formulated some far-reaching conclusions. He states that earthquakes "have their origin under ground" (Michell, 1760, p. 569), connects them with volcanism and faulting, and points out that "the motion of the earth in earthquakes is partly tremulous, and partly propagated by waves, which succeed one another sometimes at larger and sometimes at smaller distances" (Michell, 1760, pp. 571–572). He estimates the velocity of the waves in the Lisbon earthquake to have been "more than 20 miles per minute" (Michell, 1760, p. 574), and proposes a method of determining "the place of the origin of a particular earthquake from the time of its arrival at different places" (Michell, 1760, p. 626).

He believes that earthquakes are produced when subterranean fire is invaded by water, but admits that he does not know how the fire could be sustained in the absence of air, as in submarine volcanoes. He is aware of the continuity of strata and their sequence and explains how mountains are built by folding the layers through the force of subterranean heat. He considers the "compressibility and elasticity of the earth" and points out that they are "qualities which don't show themselves in any great degree in common instances, and therefore are not commonly attended to" (Michell, 1760, p. 597). "This compression must be propagated on account of the elasticity of the earth, in the same manner as a pulse is propagated through the air" (Michell, 1760, p. 599). He also observed that earthquake waves at sea (now called tsunamis) travel at velocities that depend on the depth of the water.

It was believed in his time that volcanism and earthquakes were restricted to the proximity of the sea, that pyrite may be the fuel that powers volcanoes, that earthquakes are related to the weather, and that they, in turn, affect the magnetic needle. Point by point, Michell categorically denies all those notions, but his penetrating analysis seems to have escaped his contemporaries.

The power of his intellect was respected, and his work on earthquakes surely played a part in his being elected to the Royal Society the year he presented it, and to the Woodwardian chair two years later, but its fundamental importance was not understood. His paper produced virtually no scientific response until it was noticed by the *Edinburgh Review* two generations later and reprinted in the *Philosophical Magazine* (Michell, 1818).

Late in life Michell built a torsion balance for measuring the Newtonian gravitational attraction between bodies of known masses (what is now called the gravitational constant) thus "weighing the earth." It was a brilliant idea, but Michell died before he could perform the experiment. The elegant apparatus went to his friend Henry Cavendish (1731–1810) who refined it further, made the observations with great

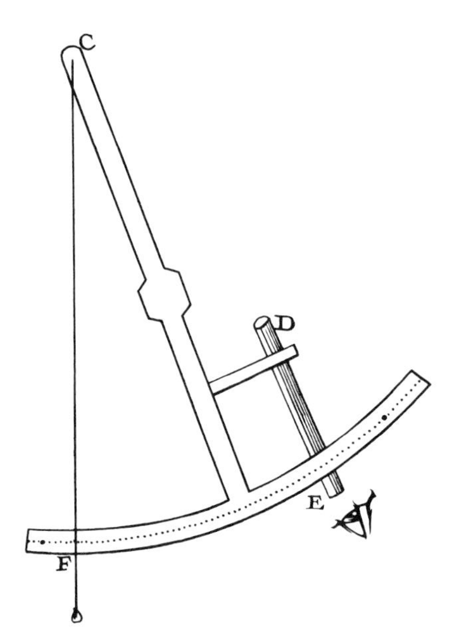

FIGURE 7.2. Schematic diagram of a zenith sector: C, center of rotation; D, telescope objective; E, eyepiece; F, plumb line. (After Bouguer, 1749)

skill, and calculated that the average density of the whole Earth was 5.48 times greater than the density of water (Cavendish, 1798). (For a succinct but full account of this research see Poynting, 1913.)

Earlier, Nevil Maskelyne (1732–1811), the Astronomer Royal, had approached the same question in another way. He was aware that mountains affect the plumb line, as predicted by Newton (1729) and discussed by Bouguer (1749), and that a quantitative determination of the effect would permit a calculation of the average density of the Earth.

FIGURE 7.3. Schiehallion Hill seen from Loch Rannoch, looking ESE.

Astronomers, like geophysicists, tend to tailor their experiments to the capabilities of their equipment, and Maskelyne knew that with the Royal Society's "zenith sector" he could measure the angle between the plumb line and the directions to fixed stars within a few seconds of arc.

For his mountain Maskelyne picked a large, steep, and roughly symmetrical ridge, striking approximately east-west (Schiehallion Hill, elevation 3547 feet, in northern Perthshire). He set up an observatory on the south slope and then another on the north slope of the mountain and meticulously determined the direction of the plumb line with respect to selected stars at both locations. In the meantime, his friends and assistants made a precise theodolite survey of the topography of Schiehallion and its vicinity (Maskelyne, 1775).

He found a difference of 11.6 inches between the directions measured at the two observatories, corrected for the curvature of the Earth, and calculated "the mean density of the earth to be double to that of the hill" (Maskelyne, 1775, p. 533). The calculations were refined by Maskelyne's friend, Charles Hutton, professor of mathematics at the Royal Military Academy at Woolwich, who concluded that "the density of common stone is to that of rain water as $2\frac{1}{2}$ to 1; which . . . results (in) the ratio of $4\frac{1}{2}$ to 1 for the ratio of the densities of the earth and rain water" (C. Hutton, 1779, p. 782). Maskelyne had a clear picture of the importance of the experiment, and a tremendous amount of careful work had gone into it, but neither he nor Hutton felt it necessary to bring in a "mineralogist" to look at the "common stone" of Schiehallion and make a determination of its density with a precision commensurate to the quality of the other measurements. The area is now known to be underlain by Dalradian quartzites, marbles, and mica schists folded into a complex structure, and with an average density more than 10 percent higher than Hutton's "common stone."

The Michell-Cavendish experiment, with its thorough control over variables, provided a more accurate determination of the mean density of the Earth (5.48, compared with the presently accepted value of 5.517), but the comparatively uncontrolled Maskelyne-Hutton determination, made as it was on a real mountain, was preferred by many geologists. Both results strongly implied the presence of some heavy and presumably metallic material within the Earth, and that remained a puzzle.

The heaviest well known rock was basalt, and its density was just a little over 3. The question of its origin had been hotly debated in continental Europe, especially in France, but created very little stir in England. In Italy, William Hamilton (1730–1803), British ambassador to the Court of Naples, became interested in volcanoes and made extensive studies of Vesuvius and Etna. He trained a good local artist, Pietro Fabri, to make elegant and scientifically accurate sketches to go with his own methodical accounts of what he had seen, and sent his observations to

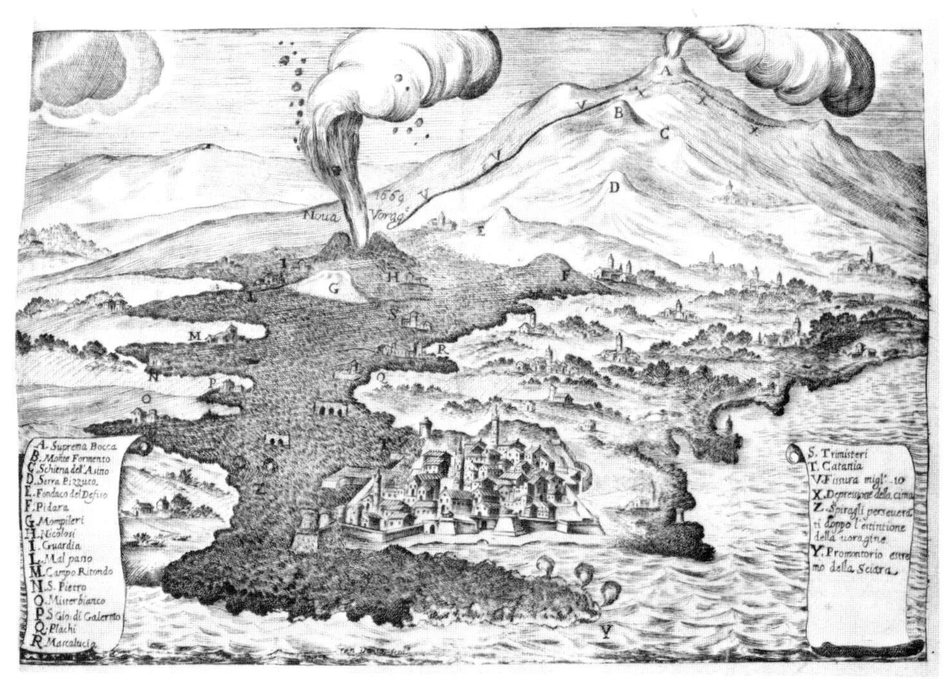

FIGURE 7.4. Eruption of Etna and the great lava flow of 1669. (Borelli, 1670; photograph courtesy of R. D. Gurney, Ltd., London)

the Royal Society. He was elected Fellow, and saw his letters published in the *Philosophical Transactions* from 1767 to 1795. These firsthand accounts of the progress of volcanic eruptions finally excited some interest in England and were reprinted several times, but their impact on British geology was not profound. Volcanism was still regarded as a curiosity and Hamilton himself considered it in that light. Just the same, his privately printed book of observations on Etna and Stromboli with a supplement on Vesuvius, beautifully illustrated by Fabri (Hamilton, 1776, 1779), remains one of the handsomest geology books of all time.

One of Hamilton's readers was John Whitehurst (1713–1788), a renowned clockmaker from Derby who was appointed official Stamper of the Money-Weights, moved to London, and wrote a book of fanciful speculation on the origin and history of the Earth (Whitehurst, 1778). The book is not important for its theories, but for Whitehurst's recognition and accurate description of the regular sequence of the strata of the Derbyshire mining region, which he had obviously studied very thoroughly. Having never seen a volcano, he recognized from Hamilton's descriptions that the massive strata of "toadstone," interbedded with the thick fossiliferous limestones, are of an igneous origin and "leave no room to doubt of its being as much a *lava* as that which flows from

Hecla, Vesuvius, or Aetna" (Whitehurst, 1778, p. 150). The "toadstones" are now described as olivine dolerite flows and sills.

Whitehurst was familiar with the French work on volcanism, but most of his English colleagues were not. The field was wide open for an uncommonly observant writer, naturalist, numismatist, and romantic opportunist recently arrived from Germany—Rudolf Erich Raspe (1737–1794). He had written a pretentious but important little book (Raspe, 1763), notable for being among the first to rediscover the value of the geological ideas of Robert Hooke, and full of exaggerated praise for him. It was dedicated to the Royal Society, with the obvious purpose of getting Raspe elected Fellow, and actually succeeded to that end in 1769.

Unfortunately, Raspe had pawned some gold coins from the collection of the Markgrave of Hesse, who had engaged him as his curator, and he had to leave Germany in a hurry when the shortage was discovered (Carswell, 1950). He fled to England and was busy crashing high society when the Royal Society found out about the coins and voted to eject him in 1775. For the rest of his life Raspe drifted, but not without leaving his mark on British geology.

Raspe had recognized the significance of those paragraphs on prismatic basalt that Desmarest had planted in the list of plates for volume VI of the *Recueil* for the *Encyclopédie* (Diderot, 1768). That led him to study the basalts around Kassel in Germany, and to discover that the Habichtswald was another region of extinct volcanoes (Raspe, 1771), the first known in Germany. Newly arrived in London, he translated his work into English and published it again, billing it "as supplementary to Sir William Hamilton's observations on the Italian volcanos" (Raspe, 1776). That was the first time English readers were told about the basalt controversy and Desmarest's work in the Auvergne, but their response, with the exception of Whitehurst and very few others, appears to have been total disinterest.

Undaunted, Raspe continued to publish translations of important new German geological works for the London market. He brought out Ferber's *Travels through Italy* (Ferber, 1776), important because it introduced Arduino's classification of rocks into Primary, Secondary, and Tertiary, and Born's *Travels through the Bannat* (Born, 1777), which gave news of the first stirrings of the chemical mineralogy that began with the early work of Urban Hiarne (1641–1724) in Stockholm and was now being developed in Germany and in Sweden by Johann Heinrich Pott (1692–1777) and Johan Gottschalk Wallerius (1709–1785). The books strengthened Raspe's tattered reputation but did not provide enough to live on. For years, off and on, he was in the service of Matthew Boulton (1728–1809), the minter, manufacturer, and partner of

James Watt (1736–1819) in the steam engine business. In that connection he served as assay master at the Dolcoath Mine near Camborne in Cornwall.

Later, Raspe moved to Edinburgh and met Hutton and Black, but not much is known of any geological interactions among them, although it is established that Black treated him for lameness (Carswell, 1950). He became immortal in literature quite unintentionally at Dolcoath, when he tossed off a fancifully humorous little pamphlet that he later sold to a publisher in London for a pittance. It appeared anonymously in 1786 as *Baron Munchausen's Narrative of His Marvelous Travels and Campaigns in Russia*, and became one of the greatest favorites of young and old readers of the nineteenth and early twentieth century.

MEANWHILE, BACK IN EDINBURGH

Things were looking up in Scotland in the second half of the eighteenth century. The union with England, voted in 1707, had turned out to be advantageous to the Scots. Industries were developing, agriculture was improving, and business was good. Edinburgh may have been as drab and rainy as ever, but the "Scottish Renaissance" was gaining momentum and the intellectual climate in the city was radiant.

Religious differences with England were not forgotten, and, indirectly, they also worked to the advantage of Scottish universities. Cambridge and Oxford would confer degrees only on members of the Church of England. Because the Scots were Calvinists, those among them who felt strongly about it or who did not care for the restrictive religious pretensions practiced in England stayed closer to home and went to their own enlightened University. There they were joined by a growing number of dissident Englishmen who, for various religious and political reasons, preferred the liberal spirit of the Scottish educational system.

There was also the matter of quality. By 1800, the University of Edinburgh had more students than Oxford and Cambridge combined, and the view was widely held that it was the world's leading school of medicine and science. The city of Edinburgh had less than one-tenth the population of London, and the per-capita concentration of talent there was prodigious.

Beginning about 1766, Joseph Black (1728–1799) lectured there in chemistry, having established his reputation a decade earlier when he published his experimental proof that "fixed air" (carbon dioxide) was present in soda, potash, limestone, and "magnesia alba" (hydrated magnesium carbonate). His close friend James Hutton (1726–1797) came up from Slighshouses (the Hutton family farm in Berwickshire) about 1768, and in time they became the leading lights on the Edinburgh scientific scene. Together with Adam Smith (1723–1790), the economist, they founded the Oyster Club (Playfair, 1805), a friendly "little society" that met weekly to dine, discuss advanced ideas, and arrange an occasional field trip.

Eating local oysters was a fashionable entertainment in Edinburgh at the time (Topham, 1776, p. 129), and other local worthies joined the club. When distinguished visitors appeared in Edinburgh, they were invited to dinner, and the Oyster Club became an important center of scientific communication and debate, fueled in part by the powerful porter brew that came with the oysters.

John Playfair (1748–1819) first appeared as a student in 1769 and was appointed Professor of Mathematics in 1785. Sir James Hall (1761–1832), by far the youngest of the group, first came to Black's lectures in 1781, commuting from his country seat in Dunglass, not far from the Hutton farm. He had been a classmate of Playfair's in the geology course at the university and they became close friends. His pioneer work in high-pressure high-temperature mineralogy (now called experimental petrology) began about 1797, shortly after Hutton's death, because Hutton had thought that the experiments were unnecessary and Hall respected his wishes while he was alive.

Hutton's geological ideas, developed over some three decades of thought and observation, were sharpened in many discussions with Black and other critical thinkers. They were translated into readable English by Playfair (1802), and experimentally supported by Hall (1800, 1806). The ideas set geology on its course, but it was done in a very quiet, low-key manner. A new system of geology had been created, but its fundamental importance was largely unrealized for more than a generation.

Excerpts from the *Abstract of a Dissertation* (Hutton, 1785):

The purpose of this dissertation is to form some estimate with regard to the time the globe of this Earth has existed, as a world maintaining plants and animals; to reason with regard to the changes which the earth has undergone; and to see how far an end or termination of this system of things may be perceived, from considerations of that which has already come to pass.

The solid parts of the present land appear, in general, to have been composed of the productions of the sea . . . Hence we find reason to conclude,

1st, That the land on which we rest is not simple and original, but that it is a composition, and has been formed by operation of second causes.

2dly, That, before the present land was made, there had subsisted a world composed of sea and land, in which were tides and currents, with such operations at the bottom of the sea as now take place. And,

Lastly, That, while the present land was forming at the bottom of the ocean, the former land maintained plants and animals; at least, the sea was then inhabited by animals, in a similar manner as it is at present.

Hence we are led to conclude, that the greater part of our land, if not the whole, had been produced by operations natural to this globe; . . .

By thus proceeding upon investigated principles, we are led to conclude, that, if this part of the earth which we now inhabit had been produced, in the course of time, from the materials of a former earth, we should, in the examination of our land, find data from which to reason with regard to the nature of that world, . . .

If we could measure the progress of the present land, towards its dissolution by attrition, and its submersion in the ocean, we might discover the actual duration of a former earth; . . . But, as there is not in human observation proper means for measuring the waste of land upon our globe, it is hence inferred, that we cannot estimate the duration of what we see at present, nor calculate the period at which it had begun; so that, with respect to human observation, this world has neither a beginning nor an end.

Hutton's geology rests on the concept of continuous natural processes acting over periods that are infinitely long compared with a human life span. He took most rocks to be sedimentary. Decay and erosion of the land produce the sediment and running water moves it to the sea, with floods and storms doing their part. The internal heat of the Earth, acting at high pressures, consolidates the stratified sediments and converts them into rocks. In the course of unimaginable spaces of time these rocks are raised from the bottom of the sea and in the process perhaps bent or contorted by "this agent [which] is matter actuated by extreme heat, and expanded with amazing force" (Hutton, 1788, p. 266).

"The raising up of a continent of land from the bottom of the sea is an idea that is too great to be conceived easily," he admits (Hutton, 1788, p. 295), but he saw evidence that it happened and left the explanation for the future. It was clear to him that igneous masses, dikes, and veins may intrude sedimentary rocks from below during these motions, and he insists that about granite "there is sufficient evidence of this body having been consolidated by means of fusion and in no other manner" (Hutton, 1788, p. 257). He saw in nature "wisdom, system, and consistency," but thought "it is in vain to look for anything higher in the origin of the earth. The result, therefore, of our present inquiry is, that we find no vestige of a beginning,—no prospect of an end" (Hutton, 1788, p. 304).

The work appeared in print in three progressive versions. First came the anonymous 30-page *Abstract* (Hutton, 1785), distributed privately about the time when Hutton first read his paper to the Royal Society of Edinburgh in March and April of 1785. The full text was published in the first volume of the Society's *Transactions* (Hutton, 1788) and by that time it had grown to 100 pages. The scathing comments the paper elicited from Richard Kirwan (1793) deeply annoyed Hutton and prodded him to collect more "proofs and illustrations" to greatly expand it. He brought out two volumes of the newly reinforced *Theory* in 1795 with a total of 1204 pages.

The manuscript of six additional chapters was found a century later and published by the Geological Society; it was edited by Geikie as volume 3 with 267 pages (Hutton, 1899). A collection of 70 drawings made between 1785 and 1788 by Hutton's close friend John Clerk of Eldin (1729–1812) was found in a portfolio at Penicuik House, about 9 miles south of Edinburgh, in 1968. Clerk had been Hutton's fellow traveler on many field trips, and about half the drawings apparently had been made for the final edition of the *Theory*. Twenty-seven of them have now been published in faithful facsimile (Craig, McIntire, and Waterston, 1978).

Hutton's prose was not easy to read even for his contemporaries, and the *Abstract*, in its brevity, is probably the clearest expression of his ideas. Fortunately he had a capable interpreter in John Playfair who was a skillful writer and intimately familiar with Hutton's material (Playfair, 1802). Wrote Humphry Davy (1778–1829) in 1805: "Dr. Hutton is obscure and perplexed from the multitude of facts which crowded on his mind. Mr. Playfair, gifted with the faculty of selection, has discriminated such phenomena only as are calculated to elucidate his opinions. He has given to the Plutonian theory a new, a more philosophical, and a more fascinating form" (Davy, 1980, p. 58).

The debate over the *Theory* was not particularly broad, but it was intense. The two most prominent critics were Jean André DeLuc (1727–1817), Citizen of Geneva, F.R.S., Reader to Her Majesty (Charlotte Sophia, Queen of England), internationally conspicuous at the time but now largely forgotten (except for his work in meteorology), and Richard Kirwan (?1733–1812), Irish chemist and mineralogist, one of the final defenders of the phlogiston theory, and president of the Royal Irish Academy for the last 13 years of his life. Both spent much effort on attempts to correlate geology with Genesis, but only one attacked Hutton on religious grounds (DeLuc, 1809), whereas the other adopted a coldly scientific posture (Kirwan, 1793, 1802) as if the religious issue did not particularly bother him. He denies the significance of weathering and erosion, negates the importance of heat in consolidating sediments, and cannot conceive of a substance such as granite crystallizing from a melt.

To Kirwan, Hutton's *Theory* is "fanciful and groundless." "Where then will he find," he asks, "those enormous masses of sulphur, coal, or bitumen necessary to produce that immense heat necessary for the fusion of those vast mountains of stone now existing?" Furthermore, calcareous spars are "absolutely infusible in any degree of heat yet known," and "the strongest heat that art can produce is scarcely capable of producing the slightest emollescence in pure quartz." Invoking the action of heat in such processes is "gratuitous" because there is good experimental evidence that "siliceous earth, sufficiently divided, was soluble in all acids," and even "in mere water." Apparently it was Kirwan (1794, v. 1, p. 455) who introduced "the fanciful appellations of the Plutonic and the Neptunian" to distinguish "these opposite systems" (Murray, 1802, p. 5).

Sir James Hall also drew Kirwan's fire when he announced his discovery that molten basaltic glass will crystallize if allowed to cool slowly (Hall, 1800; Kirwan, 1802), but unlike Hutton, he was not bothered. He continued with his experiments, made a pressure chamber out of an old cannon, and reached pressures close to 300 bars at temperatures up to 1000 degrees Centigrade. With that bomb he confirmed Hutton's theoretical prediction (Hutton, 1788, p. 271) that limestone would not calcine when heated under pressure but would melt and crystallize into marble upon slow cooling (Hall, 1806). These were tremendous results, but neither Playfair's prose nor Hall's experimental evidence were enough to overcome the fundamentally religious objections to the plutonist view and to stem the march into Edinburgh of Werner's neptunism.

The importance of Hutton's ideas was obvious to some, but few were prepared to follow. His changing Earth made any concordance with Genesis all but impossible, and his "indefinite space of time" left no place for the time estimates based on biblical chronology. He did write of "an order not unworthy of Divine wisdom" (1788, p. 210), but few of his readers were ready to consider that kind of abstract deism. His work was read, translated, and discussed at some length, but then it was gingerly set aside in favor of the religiously more palatable concepts emanating from Freiberg.

NEPTUNISM, EDINBURGH BRAND

In Hutton's time, geology was taught at Edinburgh by John Walker (1731–1803), a clergyman-geologist-botanist who had accumulated considerable field experience by the time he was finally appointed Professor of Natural History in 1779 and began his lectures about 1781. He published practically nothing on geology during his lifetime, but the notes for his lectures are preserved (Walker, 1966); they show that his approach was scientifically sound, but his course was orthodox to the point of being old-fashioned, even when it just began. He was a devout Presbyterian and became prominent in the Church during his academic career. His reluctance to indulge in geological theory likely stems from a fear that it could lead to conflict with religious dogma.

Walker was aware of the new ideas Raspe was introducing into Great Britain from the Continent, and he took the trouble to read Hooke

(1705), but he could not agree with him that many fossils were extinct or that earthquakes and volcanism were factors in mountain building. He noted the continuous decay of the Earth's surface but was not ready to admit that the landscape is shaped by running water. He accepted Lehmann's division of geological formations into Primary and Secondary, but took a cautious position on fossils and was convinced that species are fixed and that none could have disappeared in the past. His respect for scientific observation is evident and he clearly discerns what is known and what is not, but he shies away from theory and interpretation.

In his criticism of the Maskelyne-Hutton determination of the mean density of the Earth, for example, he shows that the value Charles Hutton assumed for the surface density of the Earth was too low, but then he dismisses the conclusion that the Earth may have a metallic core as "a mere supposition and indeed rather an improbable one" (Walker, 1966, p. 168) and rejects the results of the whole experiment.

Walker's course was not required in any curriculum, but it was popular and well attended (Scott, in Walker, 1966). Most years he had more than 50 paying students, and the roster of November 1782, includes the names of "Sir James Hall of Dunglass Bart" and "The Rev. Mr. John Playfair." Hall was 21 and Playfair 34 when they took Walker's course. Both were learning their geological basics from him, but neither adopted his conservative outlook. Eleven years later, Robert Jameson (1774–1854), aged 18, took the course and became a devoted disciple (Sweet and Waterston, 1967).

Hutton and Walker certainly knew each other well, and Walker, as one of the Secretaries of the Royal Society of Edinburgh, almost certainly had been present when Hutton read his famous paper in 1785. The substance of Walker's course must have been similar to the material Werner had been exposed to as a student in Leipzig a few years before, and Walker must have heard of his mineralogy, but neither Hutton nor Werner are mentioned in the lecture notes. The tenor of the notes makes it easy to guess, however, that given a choice of the two theoretical systems, Walker would have learned toward Werner's.

Thus it may not be surprising that Jameson became one of Hutton's most persistent critics and a confirmed Wernerian early in his student days (Sweet and Waterston, 1967). He even went to Freiberg for a year in 1800 to study with the master himself. In 1805 he published *A Treatise on the External Characters of Minerals,* which closely followed the style and the substance of Werner's example and established him as the major Wernerian in Great Britain. A year before, with much help from Kirwan, he had been appointed to succeed Walker in the chair of natural history at Edinburgh and there he remained for almost exactly 50 years (Sweet, 1969), broadcasting the Edinburgh brand of neptunist geognosy.

In the period of about 1800 to 1820, the University of Edinburgh had over 1500 matriculated students, and the *Bergakademie* Freiberg less than 100 (J. Eyles, 1973, p. 70). Jameson's lectures were unspectacular but interesting and popular. Thus it came about that barely a decade after Hutton's death Edinburgh became a fountain of neptunism and many more Wernerians were trained there under Jameson than in Freiberg under Werner himself.

The intellectual quality of Jameson's teaching was high and it was not the custom at Edinburgh to discourage student opinion. Many students accepted the Wernerian dogma, but a few were unimpressed and one of them, William Fitton (1780–1861), wrote a brilliant critique of the Wernerian system that did some damage to the cult of the Freiberg oracle (Fitton, 1813). Possibly out of modesty, but more likely from caution, he signed it only "F" and waited three years after graduating as an MD before he published it. Much later he gave up his medical practice and became a prominent geologist in London.

In addition to his professorship, Jameson was charged with the job of keeping the small museum of natural history that Walker had struggled so hard to maintain (without official support), and he did his best to expand it into something respectable (Sweet, 1972). James Hutton had died intestate, but through the efforts of Joseph Black and the Royal Society of Edinburgh his rock collection had been given to the university museum while Walker was still alive. Consisting mainly of samples of "common rock," some of them quite large, it was not the kind of collection that would have appeared valuable to a museum of natural history at the time. By the time the university's museum was transferred to the Crown after Jameson's death, its collections had been enlarged enough to form a good portion of what is now the Royal Scottish Museum, but Hutton's rocks were no longer there.

Over the years, more and more Wernerian concepts evaporated under the weight of geological evidence, and long before the end of Jameson's tenure little more than the bare formalism remained of Werner's geognosy. Scotland continued to produce far more geologists than one would expect from such a small population, but the Edinburgh renaissance had run its course. Jameson was still in the middle of his career when the spotlight on British geology shifted from Edinburgh back to London, Oxford, and Cambridge.

ENTER THE PROFESSIONAL GEOLOGIST

The steam engines of Watt and Boulton were pumping and hoisting in the mines, and beginning to drive machinery in factories. They were powerful and turned faster than windmills and waterwheels, but they

FIGURE 7.5. Restored hoisting engine of the East Pool Mine, between Camborne and Redruth, Cornwall. The steam engine, built in 1887 in Camborne, is a late specimen of the type that pumped and hoisted in Cornish mines for a century.

consumed a lot of coal—far more than could be hauled by horse and wagon over country roads. In the technological context of the time, the most economical solution was barges and the canals to float them on. To build canals one had to have accurate surveys and some geological knowledge. The digging itself produced new exposures in a country covered almost entirely by vegetation, and this generated more geological interest, especially where the diggers uncovered fossils. Thus it may not be surprising that the next great advance in geology was made by an English canal engineer.

William Smith (1769–1839) was neither doctor, parson, nor educated gentleman of leisure, and he was never elected Fellow of the Royal Society, but he was an early specimen of a new breed among geologists: the professional practitioner, living by his geological work, and doing well. He began as a land surveyor, mapped some coal mines, and became interested in the sequence of strata. He was involved with the Somerset Coal Canal from about 1793 until 1799, became thoroughly familiar with the stratigraphy of the region, and about halfway through the job made the important discovery that limestones which otherwise looked very similar could be distinguished by the characteristic fossils found in them (J. Eyles, 1969*ab*).

This was the key to geologic mapping, and about 1799 Smith drew his first map, showing the sequence and distribution of strata in the neighborhood of Bath. More manuscript maps followed, but none were published because by that time Smith was a famous engineer. His services were in great demand, and he felt no urge to rush into print. His geologic map of England and Wales, on a scale of 5 miles to the inch, only appeared in 1815, with a new feature that seemed necessary at the time: a darker shade of color at the base of each formation, to show the time sequence of the strata. The geology of England, with its great succession of relatively undisturbed sedimentary formations, was ideal for the development of what we now call biostratigraphy, and the cartographic beauty and accuracy of Smith's maps were a powerful beginning to that development.

The Industrial Revolution was now well under way and the demand for geological expertise was rising apace, but the great English universities took little notice. At Oxford, John Kidd, MD (1775–1851) became Reader in Chemistry in 1801, beginning a long career in the sciences including mineralogy and geology (Kidd, 1809, 1815). He was impressed by Haüy's chemical and crystallographic approach to mineralogy and was the first British mineralogist to follow Haüy's classification. In Cambridge, Edward Daniel Clarke (1769–1822) began lecturing in mineralogy in 1807 and was so well received that a special professorship was arranged for him a year later, or so it is said (Otter, 1824). He had studied an eruption of Vesuvius with Hamilton in 1792 and 1793, and traveled widely, collecting rocks, minerals, antiquities, and coins. Then he settled down to his fashionable lectures and wrote a six-volume book of *Travels* that sold well (Clarke, 1810–1823). These were the first manifestations of the scattered but growing interest in geological subjects at the great English universities.

The intellectual power of Edinburgh continued and the movement at Oxford and Cambridge was beginning, but the heart of the Empire was in London, which was where geology now found its most fertile ground. Industrial development was a powerful stimulus. Landowners were looking for coal and soon discovered that one had to know some stratigraphy to guide the prospecting. Roads and canals were being planned and the engineers were learning that geology could be a big help in such projects.

The steam engine fostered technological development in all areas, including better and cheaper methods of printing. Newspapers were proliferating, books were becoming less expensive, and public interest and participation in all this development were growing fast. Inevitably, the intellectual territory that had traditionally belonged to theology was slowly being invaded by discussions arising from the new technology. Mechanical power was here to stay, but the clergy were slow to recognize this. Unable to give new answers to the new questions, they now began to lose their monopoly on the leadership of intellectual development. The word of Moses was being drowned out by the clatter of the steam engine.

Intellectual entertainment was being found in scientific lectures, and the word was out that the best lectures and the most exciting demonstrations in London were to be seen and heard at the Royal Institution on Albemarle Street. The place had been founded in 1799 by the American-born physicist Benjamin Thompson, Count Rumford (1753–1814), as a science-teaching museum, and it was still settling down when the Managers hired Humphry Davy (1778–1829) to lecture in chemistry (and later in geology). Davy was only 23, but within a year or two he showed them that they had made a wise choice. Most of the institution's income derived from the fees paid by subscribers to these lectures, and it was not long before the charming young man was also a major fiscal asset because his lecture hall was always packed.

Manuscripts of the 10-lecture series in geology, which Davy gave beginning in 1805, have survived (Davy, 1980), and they give good insight into the geological thinking of a very promising young chemist at a time when Playfair's transcription of Hutton's geology was still new, and when Jameson's influence as a teacher of the Wernerian viewpoint was just beginning to be felt.

Davy prepared his lectures with care. His notes show that he reviewed the history of the science as found in the classics, surveyed the English geological literature, and became familiar with the progress made on the Continent. He had seen a fair amount of geology in the field, especially in his native Cornwall, and in 1804 he took a long trip to Scotland to visit important localities and collect minerals. He met John Playfair and Sir James Hall, and they took him to see some of the

FIGURE 7.6. The unconformity at Siccar Point, near Cockburnspath, 37 miles east of Edinburgh.

places Hutton had cited as illustrations for his *Theory*, including the angular unconformity at Siccar Point, about 40 miles east of Edinburgh, where gently-dipping Late Devonian sandstone overlies vertical Late-Silurian graywackes in a spectacular exposure.

Brilliant chemical craftsman that he later turned out to be, young Davy appears to have been an indifferent field observer. One can surmise that Hall and Playfair did not miss any chance of expounding the Huttonian position, but the sketch Davy made at Siccar Point (Davy, 1980, p. xxiv), perhaps even while they watched, implies that he missed the significance of what they were trying to show him. Intellectually he understood the principles of Hutton's geology and the meaning of Hall's experiments with slowly-cooling melts, but emotionally he was not ready to abandon the conventions of Genesis, the teleological argument, and Bishop Ussher's time scale. Davy never obviously bent his geology to comply with Genesis, but even much later, in his last work (Davy, 1830) written after he had seen the volcanoes and lava-filled valleys of the Auvergne, he still firmly refused to accept Hutton's continuous processes and infinity of time.

Davy's lectures were not intended for professionals, but they provided a focal point for the gathering of a few gentlemen whose interest in geology was more than casual. Most of them were mineral collectors who had founded the British Mineralogical Society in 1799 in order to share their interest, but now some of them were looking beyond their mineral cabinets and wanted to foster scientific inquiry and the collection of geological data. Some had seen Smith's manuscript maps and realized their significance; others were responding to the spirit of the times. All agreed that geological studies were more than a pastime.

THE GEOLOGICAL SOCIETY

On November 13, 1807, these gentlemen met in Freemason's Tavern on Great Queen Street near Kingsway, where they founded the Geological Society as a scientific organization that was to maintain a library and a study collection, discuss and disseminate geological observations, and "adopt one nomenclature." If the ghost of Guettard happened to be there that evening, he must have enjoyed watching his latter-day English colleagues rediscover his own ideas of more than 50 years ago.

George Bellas Greenough (1778–1855), gentleman geologist, was the Society's first president (1807 to 1813), and he was mainly responsible for its success. He quickly assembled a respectable geological library and map collection, mostly through his own gifts and those of other members, and put together the beginnings of a study collection. He hired Thomas Webster (1773–1844), the young architect of the Royal

Institution building, to take care of the museum, help with the drafting and illustrations, and generally run the Society.

This trend was somewhat upsetting to Sir Joseph Banks, the rich and energetic long-time president of the Royal Society. Sir Joseph was not opposed to the organizing of the Geological Society, but he did feel the sting of competition and would have preferred to have the geologists operating under the wing of the Royal Society in some fashion. Greenough and most of his friends already were Fellows of the Royal Society, and some of them, especially Davy, thought Banks was too political and autocratic. Furthermore, many found the Royal Society too social and not scientific enough. As always, many Fellows were being elected for their accomplishments in science, but even more were chosen for their social position and financial influence, and it is true that under Banks' presidency the Council was often dominated by people with only slight interest in science.

The geologists were serious about their science, determined to maintain standards, and beginning to be concerned with professional recognition. Greenough fought off Sir Joseph without losing a friend, it is said (Woodward, 1908), and the Geological Society retained its independence. By April 1811, it had 200 members and was becoming a major factor in the accelerating development of British geology.

Besides Davy and Greenough, several other prominent names appear on the list of early members. William Babington (1756–1833) was a practicing physician and a mineral collector who established a classification scheme for minerals based on their chemical composition (Babington, 1799). It was not revolutionary, but an obvious advance over Werner's. Jacques-Louis Comte de Bournon (1751–1825) was a geologist and mineralogist (Bournon, 1785, 1808*ab*, 1817), and a loyal defender of the Abbé Haüy. He was a staunch royalist, had fled the Revolution, and while waiting out Napoleon in London, had acted as consulting mineralogist to several major collectors and as Foreign Secretary to the Society. Richard Chenevix (1774–1830) was a skillful analytical chemist and mineralogist, another strong supporter of Haüy, and a contentious opponent of Werner (Chenevix, 1808, 1811).

John Macculloch (1773–1835) was a chemist, artist, physician, and geologist with an abrasive personality and an extraordinarily wide range of accomplishments (Macculloch, 1819, 1821, 1831, 1836). Much of his geological work was done while he was in government service, and thus he may be considered Britain's first official survey geologist.

James Parkinson (1755–1824) was a renowned physician ("Parkinson's Disease"), fossil collector, and descriptive paleontologist (Parkinson, 1804–1811, 1822). William Phillips (1775–1828) had inherited a good printing and publishing business, but his first love was mineralogy and geology. It seemed only natural that he should become printer to the Geological Society. He also wrote and published several successful textbooks (Phillips, 1815, 1816, 1818; Phillips and Conybeare, 1822).

William Hyde Wollaston (1766–1828) was a prominent microchemist, theoretical chemist, crystallographer, physiologist, and inventor of powder metallurgy. He designed the first accurate reflection goniometer capable of measuring the angles between faces of small crystals, and passed it on to William Phillips, who used it for his marvelously precise constructions of crystal habits, first published in the *Transactions* of the Geological Society (Phillips, 1811, 1814). The first important conclusion made from these measurements was that Haüy's crystal theory was oversimplified.

Wollaston joined the Geological Society fairly late, in 1812, and was active on the Council, even though his time was taken up by a myriad of technical and scientific activities, not the least of which was the booming platinum-ware business he developed on the basis of his own discoveries in the chemistry of the metals of the platinum group. It made him rich, and late in life he invested £1000 as the first major endowment for the benefit of the Geological Society (Woodward, 1908).

The *Transactions* of the Geological Society first appeared in 1811, handsomely printed and carrying 18 articles. The second volume followed three years later, and a total of five volumes with 119 papers was available by 1821. John Macculloch's name appears more than any other among the authors in the first few volumes, but unlike many other scientific-society publications, the *Transactions* were not dominated by any one person and they give a good cross section of the geological interests and positions of the members and their friends. The range was wide.

The mineralogists, almost to a man, followed Haüy's approach even if they occasionally criticized some of his conclusions. Most of the geologists were diluvialists, especially in matters connected with valley erosion, and leaned toward Wernerian classifications even though none approached the neptunistic rigidity of Jameson. Macculloch, for example, used a modified Wernerian stratigraphy but was impressed by the cooling experiments of Sir James Hall and Gregory Watt (1772–1804) (Watt, 1804), and was ready to accept the igneous origin of granite (Macculloch, 1814). Admittedly, the Society took a dim view of theorizing, but when one searches the early volumes of the *Transactions* for the theoretical views of the members, one finds many shades of opinion but no trace of even a mild Huttonian. As late as 1839, the Geological Society's library still did not have a copy of Hutton's *Theory of the Earth, with Proofs and Illustrations,* and neither did the Library of the Royal Society (Fitton, 1839, p. 455).

The funds to pay for everything the Society did in its early years came from its members, and the dues were not small. Members were expected to be affluent, and it is perhaps not surprising that their interest in geological strata was not great enough to displace the usual concerns with social stratification. Beginning in its first year, the Society made it a point to invite some well-known (and many now not-so-well-known) geologists from out of town to become honorary members, and in the early years about half the membership was honorary. Right there in London, however, they neglected to invite two men who were known well enough for their geology, but presumably failed to satisfy the social-stratum requirement: Robert Bakewell (1768–1843) and, even more conspicuously, William Smith.

Bakewell was making a living as a surveyor and consulting geologist. He gave lectures that were well received, and wrote a textbook of geology (Bakewell, 1813) that may not have been terribly profound but enjoyed many editions, including three in the United States edited by Benjamin Silliman. His influence on geological education was not trifling, and he even published a paper in the Society's *Transactions* (v. 2, pp. 282–285, 1814), but he was never admitted to membership. George Greenough, the Society's president, was acquainted with William Smith at least as early as 1808 (Woodward, 1908, p. 17) and surely knew about his maps, but he was thinking of publishing his own map and may have considered Smith unwelcome competition. Smith remained outside until 1831, when the Society finally recognized that oversight and gave him the first Wollaston Medal.

Through his insight and enthusiasm, Professor Kidd at Oxford was able to attract a group of very bright young students over the years, and they included three who became important geologists: William Buckland (1784–1856) and the Conybeare brothers, John Josias (1779–1824) and William Daniel (1787–1857). They in turn formed the nucleus of a lively and far-ranging geological study group that came to include the paleontologist William John Broderip (1789–1859) and the chemist-volcanologist Charles Bridle Daubeny (1795–1867). The names of Kidd and the elder Conybeare appear on the first list of honorary members of the Geological Society in 1807, and several London geologists, notably Greenough, Phillips, and later Roderick Impey Murchison (1792–1871), actively collaborated with the Oxford people in the studies that ultimately made Oxford a world center for geological education and research.

William Conybeare was an important factor in the blossoming of British geology. He is now known for his descriptions of the marvelous saurians from the Lias of Lyme Regis in Dorsetshire and for his highly influential textbook of stratigraphy (Conybeare and Phillips, 1822), which became a basic field manual for geologists. Among the textbooks written and printed by William Phillips had been a pocket-sized stratigraphic handbook based largely on the ideas of William Smith and reporting many of his observations, together with a stratigraphic table compiled by Buckland (Phillips, 1818).

Four years later, Conybeare rewrote, updated, and greatly expanded the book with some help from Buckland and Greenough, and made it into a remarkably lucid and thorough summary of English stratigraphy, beginning with the most recent sediments and going back to the Old Red Sandstone (later placed in the Devonian). A second volume was intended to cover the older formations, but that stratigraphy took longer to work out than Conybeare had anticipated, and the second volume never appeared. In spite of that omission, *Conybeare and Phillips* became one of the most influential works in British geology and was a standard reference for two decades of that dynamic period. It established a system and a terminology that became the basis of further development of stratigraphy on both sides of the Atlantic.

Daubeny practiced medicine for some time and then became successively professor of chemistry, botany, and rural economy at Oxford, holding the three chairs simultaneously. In geology he is remembered for his study of volcanoes and his efforts to explain volcanic heat chemically—as the result of the exothermic reaction of intruding surface water with metallic calcium and sodium he thought might be present in large quantities at depth (Daubeny, 1826). In 1837–1838, Daubeny toured the United States and Canada for a year and wrote a summary of the geology of North America (Daubeny, 1839), which gives interesting glimpses of the ideas of American geologists at the time when the state surveys were just beginning and the geological picture of the eastern United States was still hazy.

In 1820, Greenough published his own geological map of England and Wales in six sheets at 6 miles to the inch (Greenough, 1819). In his fieldwork he had received much help from Buckland and William Conybeare and in turn made his extensive field notes available to Conybeare for his famous manual (Conybeare and Phillips, 1822). All these people influenced each other and their geological careers crossed at many points, but their leader was unquestionably Buckland. He was the last great diluvialist, but the label does not do him justice. He was an excellent field observer and a perspicaceous interpreter of geological data with an unusually broad grasp of geological relationships. He had been a Reader in Mineralogy at Oxford since 1813, and when he was elected to the newly created Readership in Geology six years later, he chose for the subject of his inaugural address a defense of the Noachian Deluge (Buckland, 1820). Probably in consequence of that expression he was invited to participate in the great writing project endowed by Francis Henry, the eighth Earl of Bridgewater, with the purpose of showing "the Power, Wisdom, and Goodness of God, as manifested in the Creation."

Buckland wrote the sixth of the eight treatises, *Geology and Mineralogy Considered with Reference to Natural Theology* (Buckland, 1836), and it is perhaps the last major geological textbook written from that point of view. His most important scientific work describes the rich variety of fossil vertebrates he found in Kirkdale Cave in Yorkshire (Buckland, 1823); it was widely read and earned him the Copley Medal of the Royal Society. The discovery of these bones of large extinct mammals in England alongside the bones of species still living in Northern Europe revived interest in the question of climatic change and pointed out the importance of serious study of vertebrate paleontology. The bones also fostered Buckland's long friendship with Georges Cuvier (1769–1832) and served as an important link between English and French geology.

The little port of Lyme Regis on the west coast of Dorset was then becoming a fashionable seaside resort where "fossiling" was one of the attractions. It lies in the midst of a long line of sea cliffs exposing a splendid profile of fossiliferous Jurassic rocks, with some particularly rich layers accessible on a large tidal flat right next to the town. Henry Thomas De la Beche (1796–1855) moved there with his mother when he was still a boy and there he first became interested in geology (V. Eyles, 1971, pp. 9–11). They were a family of means and Henry trained for a military career, but finding himself in straitened circumstances when the income from his parental estate in Jamaica unexpectedly diminished, he undertook a geological survey of Devonshire for the government in 1832. Then he proposed, organized, and became the first Director of the Geological Survey of Great Britain in 1835.

About 1805, John Philpot, a London solicitor, took a house in Lyme Regis to provide for his unmarried sisters Mary (1777–1838), Margaret (?–1845), and Elisabeth (1780–1857), and they stayed there for the rest of their lives (Edmonds, 1978). What else was there for educated ladies to do in Lyme Regis but collect fossils? They assembled a magnificent

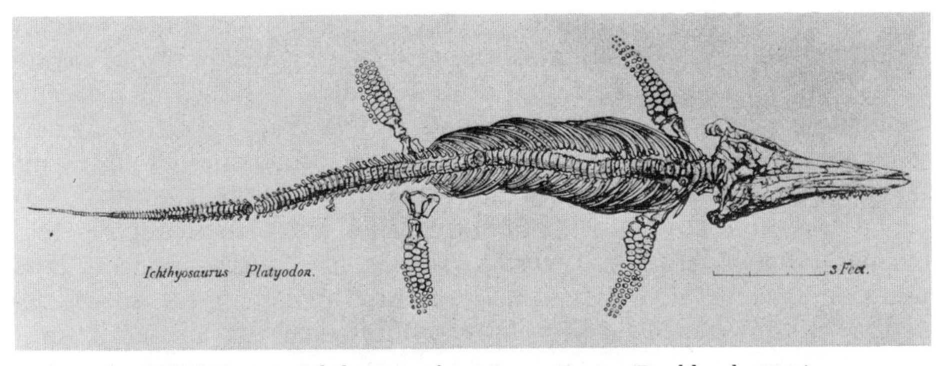

FIGURE 7.7. Ichthyosaur from Lyme Regis. (Buckland, 1836)

collection, were soon discovered by Buckland and his cohorts, and eagerly progressed under their encouragement. "The house of the Misses Philpot" became a mecca for paleontologists and fossil collectors from near and far and the collection finally went to Oxford, only after having generated much interest in geology.

Mary Anning (1800–1847), a young friend of the Philpot sisters, became a professional fossil collector in Lyme Regis and also developed a wide following. It is said that she found her first *Ichthyosaurus* on the tidal flat when she was less than 12 years old, and it created a sensation on the London geological scene. Her nearly perfect skeletons, some of them 20 feet long, made spectacular exhibits and the public went wild over the "sea dragons." Fossil bones were nothing new, of course, and serious geologists were familiar with Cuvier's recent work on the quadrupeds of the Paris basin (Cuvier, 1812*a*), but most of his bones were disarticulated and nothing like these magnificent monsters, all laid out in the rock with every limb in place.

The *Ichthyosauri* and *Plesiosauri* were described scientifically by William Conybeare (Conybeare and De la Beche, 1821; Conybeare, 1822) and eventually appeared in magnificent lithographic illustrations published by the avid fossil collector Thomas Hawkins (1810–1889)—accompanied by some of the gushiest prose ever found in the paleontological literature (Hawkins, 1840).

Another great fossil hunter, perhaps the most diligent of them all, was Gideon Algernon Mantell, MD (1790–1852), practicing physician (more or less), early friend of Lyell, and author of an important monograph on Cretaceous and Tertiary fossils from Sussex (Mantell, 1822). About that time he was fossiling with his wife around the quarries of Tilgate Forest in the Wealds of Sussex, when she noticed a group of large fossil teeth by the roadside; they turned out to be unlike any seen before. After much searching, Mantell discovered that the teeth were similar to the dentition of the modern iguana, only very much larger, and he described them as *Iguanodon* (Mantell, 1825). That was the first scientific presentation of one of the large group of giant dryland reptiles later described by Richard Owen (1804–1892) under the name *Dinosauria*.

In time, Mantell's fossil collecting got out of hand. His books were selling, but his medical practice declined, and his house in Brighton was jammed with specimens. Fossilists and gawkers kept coming and going, and finally his wife decided that she had had enough and, with the children, she moved out (Mantell, 1940).

The best geological illustrators of that time, and also great collectors and naturalists, were the Sowerbys—James (1757–1822) and his sons James de Carle (1787–1871) and George Brettingham (1788–1854). Besides a vast number of excellent illustrations of living plants, they produced two collections of fine colored plates of mineral specimens (Sowerby, 1804–1817) and a picture book of British fossil shells (Sowerby and Sowerby, 1812–1846) that was widely used in its time as an aid in identifying fossils and was a great factor in stabilizing the nomenclature.

CAMBRIDGE

Geology was slow to get started at Cambridge. Edward Clarke, the once popular lecturer in mineralogy, puttered along, and his colleagues John Hailstone (1759–1847), the Woodwardian Professor of Geology, added to the old collection and occasionally showed it to visitors until 1818, when he resigned to be married. He had studied with Werner and published a syllabus in 1792, but it is said that in his 30-year tenure of the Woodwardian Chair he gave no lectures at all (Clark and Hughes, 1890).

With the notable exception of Michell's brief incumbency, "the successive Woodwardian Professors had done little or nothing to justify their appointment. To this discreditable state of things the University not unnaturally wished to put an end" (Clark and Hughes, 1890, v. I, p. 159). Conybeare's friend, Adam Sedgwick (1785–1873), was elected to succeed Hailstone not because of any conspicuous qualifications in geology, but for his dynamic energy and for the promise that he would do something. It was hardly a promotion for him. He had been teaching at Trinity College and preaching on Sundays, and all that brought in much more than the stipend as Woodwardian Professor, which was then only £100 per annum, but he wanted independence and here was his chance.

"On Hailstone's resignation there was evidently a feeling in the University that Woodward's bequest had not produced the results which might have been anticipated" (Clark and Hughes, 1890, p. 197), and the syndics wisely recognized the root of the problem when they recommended to the Senate that the space available to the Woodwardian Professor be greatly increased and his salary be doubled. That made some difference, even though £200 was still barely enough to live on. Young Lyell's allowance from his father at that time was more than twice that, and William Smith was then earning about ten times as much (J. Eyles, 1969*a*).

Sedgwick's appointment was the turning point for Cambridge. If he knew little geology when he was elected, he learned it very fast. "Hitherto I have never turned a stone," he is supposed to have said (ibid.), "henceforth I will leave no stone unturned." And so he did. He stayed at Trinity College all his life, and his geology course continued for more than half a century. At the same time he conducted thorough and imaginative stratigraphic field studies, mainly in the oldest and most complicated fossiliferous rocks of England and Wales, which he called Cambrian after the Latinized Welsh name for Wales. He was also conspicuous in University affairs, greatly expanded the collections, and effectively founded what is now the Cambridge School of Geology.

Clarke's successor in mineralogy was Sedgwick's friend and former student, John Henslow (1796–1861), who held the chair for three years and then switched to botany. One of their students was a young man whom few regarded as promising, but who had a certain quality that Henslow appreciated. His name was Charles Darwin (1809–1882), and it was Henslow who recommended him to the Admiralty for a berth in the *Beagle*.

Earlier it was intimated that this was the era of the professional geologist, and it was, in retrospect, but the actual number of them was very small. The beginning of professional geology in England was more of a trickle than a tidal wave. Of the names mentioned in this chapter, only four would qualify as geological professionals before 1830: Smith, Bakewell, Macculloch, and Anning. De la Beche and Murchison became professional later, but by 1830 neither of them had yet earned a farthing by their geology.

English geological education was entirely in the hands of the clergy. Kidd, Buckland, the Conybeares, Clarke, Hailstone, Sedgwick, and Henslow were all men of the cloth, deriving incomes from offices in the Church of England, even though most of them spent far more time doing geology than serving the Church.

The exuberance of Buckland and Sedgwick infected their students and friends, but all their geological activities made hardly a dent in the educational outlooks of Oxford and Cambridge. The great universities paid scant attention to "the temper and spirit of these times" (Lyell, 1827*a*, p. 268). They remained unimpressed by all that digging, mapping, and fossil collecting, and continued to regard themselves as the exclusive training ground for clergymen, lawyers, civil servants, and physicians—in that order of importance. Emphasis was on the classics, with a glance or two at mathematics (at Cambridge), and geology was no more than one of many available sidelines, "optional and extracurricular" (Rudwick, 1975). When Buckland left Oxford in 1845 to become Dean of Westminster, he was glad to move because he felt that his long effort to bring natural science to Oxford had been a failure. In all of England there was still no place where a young man might get a degree in geology.

The geological legwork was done by physicians and ladies and gentlemen of leisure when the spirit moved them. Geologizing was fun for both clergy and laymen, and it was fashionable. They had the time (and usually also the money) to organize their collections, to travel, and to publish their findings. The period 1800 to 1830 in English geology was not the time of any dazzling solo performance. There was no prophet leading the way and no salient principle being established. There were just a few dozen self-taught but reasonably qualified and passionately interested people collecting geological information and slowly building up a pool of knowledge. In their midst was a student of Buckland, a young would-be barrister, an Anglified Scottish gentleman, who was destined to become the attorney and the voice of geology. His name was Charles Lyell (1797–1875).

CHAPTER EIGHT

Geology Comes of Age in Britain

SEDIMENTS AND VOLCANOES

The unifying concept that tied geology together was the development of the English stratigraphic sequence based on the correlation of strata by fossils. Many ideas had come from the Continent, but England was the obvious place for stratigraphy to flourish. From Wales to East Anglia the sedimentary rocks dip gently to the east and range in age from very young to Sedgwick's Cambrian with only minor gaps. Exploring and mapping this favorable testing ground, English geologists saw Steno's superposition take meaning, Hooke's chronology raised (Hooke, 1705, p. 411), and Hutton's unconformity become real to them—all because they had developed a perspective on sedimentary rocks. The stratigraphic table was still a crude sketch in 1830, but the enormity of geologic time was becoming obvious even to the diluvialists.

At the same time, mineralogy was advancing dramatically along the lines of the chemical-crystallographic approach brought forth by Haüy. Structural geology was growing through numerous studies of the three-dimensional picture in coal mines, and only the understanding of surface processes was noticeably lagging. Buckland was teaching that the present surface of the Earth had been shaped by the violent currents of the Deluge, Desmarest's work in the Auvergne was being ignored, and Hutton's view that valleys are made by the rivers that flow in them was considered only by a few young radicals.

Travel across the Channel was fashionable again, after Waterloo, even though France was like "a spent volcano" after "the terrific violence with which her social system had been shaken," and exhausted "after carrying desolation into all surrounding countries" (Lyell, 1827*b*, p.

FIGURE 8.1. Sir Charles Lyell, the Murchisons, Lyell's clerk George Hall, and an unidentified gentleman traveling in southern France in 1828.

437). The grand tour of Europe had always been a desirable pastime for British gentlemen, and "one might imagine Napoleon to have constructed his splendid roads for their sole use and pleasure" (Lyell, 1827*b*, p. 438). The new wrinkle of the grand tour was that quite a few Englishmen now favored the idea of studying a little geology together with all the other good things France still had to offer.

One such party leaving Paris for the south of France in May 1828 consisted of Roderick Murchison, his wife Charlotte, and their indefatigable friend Charles Lyell. Murchison had had a brief military career, become interested in geology through Davy's lectures at the Royal Institution, worked some with Buckland, and joined the Geological Society. With *Conybeare and Phillips* in his pocket he spent two summers in the field, studying strata in England and Scotland, and now was ready for geological adventure. His wife approved of his interest in geology—delighted to see him get away from the fox hunting that had taken so much of his time before.

Lyell was not yet conspicuous on the London geological scene, but neither was he unknown. He had read Bakewell's book as a boy, and taken Buckland's courses in 1817 and 1818 while he was what we would call a pre-law student at Oxford. His father wanted him to become a barrister and young Charles gave it a good try, but his heart was not in it. He took every chance he could to visit the strata he had read about in *Conybeare and Phillips*, became acquainted with more and more geologists in Britain and in France, and read his first paper to the Geological Society in 1825. For two more years he practiced a little law, but by the time he drove out of Paris with the Murchisons he was a full-time gentleman-geologist.

The previous summer he had written a long (47-page) review (Lyell, 1827b), unsigned (as was usual), of the *Memoir on the Geology of Central France*, recently published. The author of that book was his friend George Julius Poulett Scrope (1797–1876), a student of Clarke and Sedgwick, and another serious gentleman-geologist in London. Beginning in 1817, Scrope had made long trips to the volcanoes of Italy and in 1821 he spent several months in the Auvergne. He was not the first Englishman to make a thorough study of the French volcanoes, but he was probably the most observant.

Besides his careful analysis of the young freshwater sediments of the region, of their relation to the sequence of volcanic phases, and of the configuration of the landscape shaped by stream erosion and volcanism, Scrope brought back a series of magnificent panoramic sketches drawn to illustrate the geological relationships he had observed. He made it clear that the fire-belching mountains of volcanoes are not produced by catastrophic uplifts, as proposed by the "craters of elevation" hypothesis of Buch proposed in 1809, elaborated on 27 years later, and widely accepted both in England and on the Continent in the 1820s and the 1830s. Instead he concluded that volcanoes merely build up in time as rock and ashes are ejected from a volcanic vent. If the activity continues in the same place for a long time, the volcano becomes very large.

Furthermore, Scrope observed that some of the young French volcanoes had spewed lava that flowed over freshwater sediments that contain very young fossils, apparently younger than what then was and still is classified as Tertiary. Thus some of these volcanoes had to be even younger than those young fossils (Scrope, 1827). Yet they had to be older than 2000 years because Julius Caesar had marched through the Auvergne and he surely would have mentioned volcanism, had he seen any (Daubeny, 1826, p. 14). Scrope leaned toward Huttonian ideas, and the evidence of successive erosion and deposition he had seen in the Auvergne removed any doubt he might have had about the enormity of geologic time.

Lyell's review of Scrope's work on central France (1827b) is a masterpiece of writing and a clear harbinger of what was to come from him in later years. He understood the geological significance of Scrope's conclusions and recognized their power. He may have differed with the pedantic speculations about volcanism that Scrope had allowed himself in an earlier book (Scrope, 1825), but that was a different matter. The importance of what Scrope had found in the Auvergne was now very clear to him and he simply had to go and see it himself.

They spent about ten weeks in the volcanic region of central France,

FIGURE 8.2. (a) "Lava current of Thueyts near the Gueule d'Enfer," filling a valley originally cut in granite gneiss, and then cut through again by the Ardèche River. The volcano is comparatively young and "we feel disposed to ascribe almost unlimited power to ordinary rivers when a sufficient lapse of time is assumed. Nature rarely affords us, as here, an accurate measure of amount of destruction occasioned during periods of definite extent, or whose limits we can fix with reference to other natural events." (Lyell and Murchison, 1829, p. 30) *(b)* The same scene as the camera sees it.

following Scrope's leads and finding that his geology was correct almost everywhere they went. Both Lyell and Murchison were stratigraphers at heart and they paid particular attention to the thick sequence of freshwater (lake) sediments and their interrelation with the volcanic events. Again and again they saw strata that looked superficially like the classic units they knew in England (the Old Red Sandstone, the Millstone Grit, the New Red Sandstone), but on closer examination it always turned out to be much younger. One had to look closely at the fossils, not just the appearance of the rock. Scrope had observed that the oldest of these young lake beds contained no volcanic debris, so that the lake had to have been there long before volcanic eruptions began. The whole complicated history of erosion and deposition, punctuated by repeated volcanic events, had occurred since the beginning of the Tertiary—in a time span that is very short in the history of the Earth. Stratigraphy "forces us to make almost unlimited drafts upon antiquity," was the way Scrope had put it (1827, p. 165), and Lyell and Murchison agreed completely.

They went into Italy, as far as Padua, and there the Murchisons turned back, but Lyell continued southward even though he had originally planned to be back in Scotland by that time for a bit of shooting. Now he was sure he had to see the king of European volcanoes, the mighty Etna. He was delayed a week or so in Naples, saw Vesuvius and all the other sights in the vicinity, including those three columns remaining of the "Temple of Jupiter Serapis" near Pozzuoli. Many visitors, including Daubeny (1826, pp. 161–164), had speculated about the marine clam borings high up on those columns, and Lyell was no less puzzled, but he made a careful examination of the shore terraces in the vicinity and became convinced that the "temple" had sunk beneath the sea since it was built by the Romans and later was uplifted again. He made it his prime exhibit of the changeability of land and sea, and in time it became famous as the trademark illustration for the book that was then germinating in his mind (Lyell, 1830).

In late November he reached Sicily, climbed Etna, and spent six weeks reconnoitering the geology of the island. In the Val di Noto under Etna's oldest lavas, he found the limestones Daubeny had mentioned (1826, p. 201) as probably young, and found that their fossil assemblage consisted almost entirely of species that looked to him like those still living in the Mediterranean. He correlated the limestones and the "blue clay" below them from the Val di Noto across the island and it became obvious to him that the tremendous volcanic edifice of Etna, almost 11,000 feet high and some 30 miles in diameter, had been piled up since Tertiary time.

Hutton had been right about the infinity of time, Scrope had confirmed it, and now Lyell was ready to become its principal advocate. He headed back to London to write and described with Murchison (Lyell and Murchison, 1829) the volcanic rocks of central France. (For an exhaustive chronicle of Lyell's activities up to 1841, see Wilson, 1972.)

THE PRINCIPLES

One particular event marks 1830 as the end of an era: It was the year when reconciliation at last became impossible and final papers were served for the divorce of geology from religion. The attorney in the case was Charles Lyell and his brief was the *Principles of Geology*, which first appeared that year.

He introduced the case in the guise of an historical review—and with finesse (Rudwick, 1970). In the first four chapters of the first volume he goes through the geological literature, from the earliest legends to "the era of living authors," subtly selecting the views that fit his plan and blending in just enough editorial comment to lead the reader toward his own ideas. There seems to be nothing controversial in this compilation, but in the end there is no question but that religion and geology are two distinct things; it is done so neatly, so smoothly, yet so decisively that no reasonable person could object. He knew there would be some uneasiness and there was, but the time had come for these things to be said.

In reviewing the reaction to the *Principles*, one has the impression that even those geologists who objected in public were privately relieved that someone at last had made the statement. Almost overnight, the *Principles* became the book to read. Captain Fitzroy had bought a

FIGURE 8.3. Hob-nailed field shoes "presented by Professor Sedgwick, Sept. 24, 1827. Postdiluvian?" (Kinnordy House; courtesy of Lady Lyell)

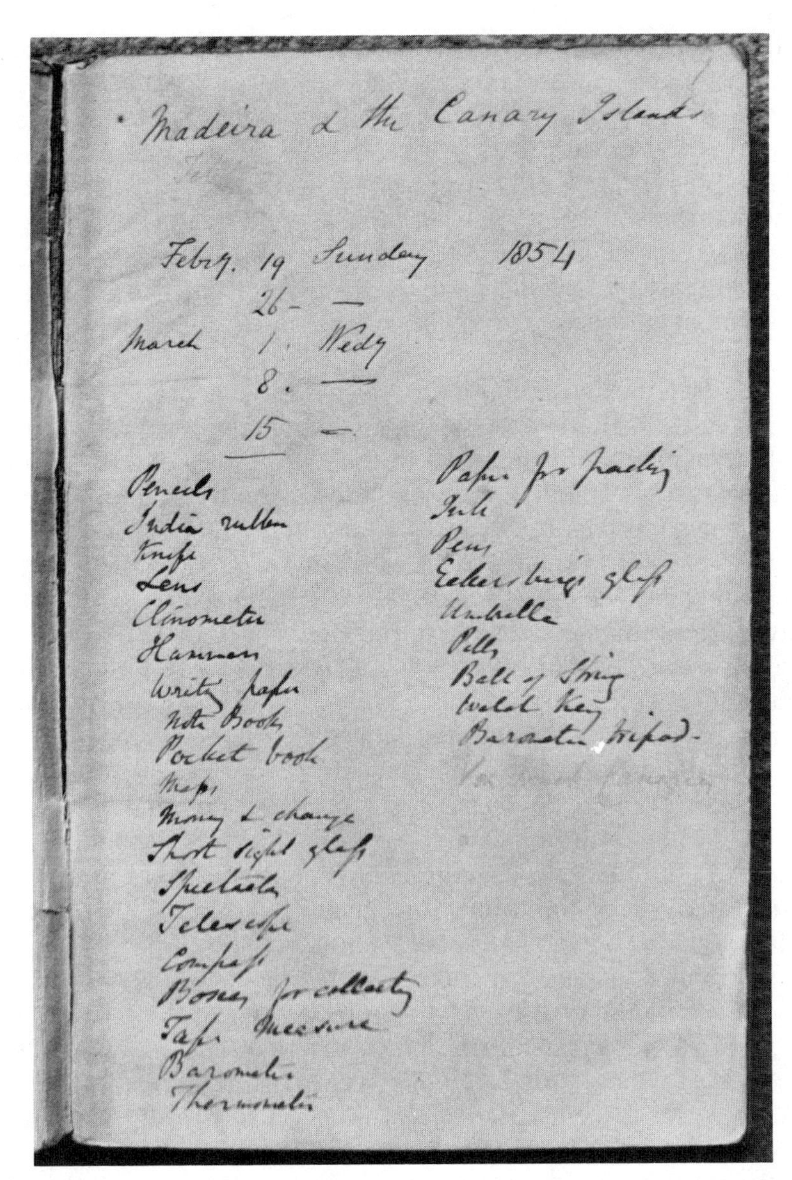
Madeira & the Canary Islands

Febry. 19 Sunday 1854
26 —
March 1. Wedy
8. —
15 —

Pencils
India rubber
Knife
Lens
Clinometer
Hammers
Writing paper
Note Books
Pocket book
Maps
Money & change
Short sight glass
Spectacles
Telescope
Compass
Boxes for collecting
Tape measure
Barometer
Thermometer

Paper for packing
Ink
Pens
Eckersberg's glass
Umbrella
Pills
Ball of String
Watch key
Barometer tripod
Von Buch Canaries

FIGURE 8.4. Charles Lyell's checklist for a field trip to Madeira and the Canary Islands, listing everything he expected to need including the guidebook, "Von Buch Canaries" added in pencil. (von Buch, 1825; notebook 193, Kinnordy papers, courtesy of Lady Lyell)

copy and as HMS *Beagle* was heading into the South Atlantic, Charles Darwin was reading it.

The *Principles* made it obvious that Lyell had mastered the subjects of geology better, and on a broader base, than any of his contemporaries. He had read almost everything, traveled widely, and his fieldwork was accurate and imaginative. He knew almost everybody who was anybody in British, French, and Italian geology, and he knew where to look and when to listen. He was not much of a lecturer (Clarke, 1923, p. 114), but he had the gift of sound judgment and an uncanny talent for organizing his material and presenting his case clearly and convincingly in print. The *Principles* had five editions in seven years, and by this time Charles Lyell was the main catalyst and arbiter as well as the voice of geology during its explosive growth in the mid-nineteenth century. Lyell's own conceptual contributions to geology are dwarfed by the importance of his position as interpreter, judge, and communicator of the work of others.

The subdivision of Tertiary time based on the proportion of "still-living" species in the fossil assemblage is usually credited to him, but several other people participated in that development and it would be pointless to argue about priority. The taxonomic paleontology upon which the system was founded was largely developed by Gerard Paul Deshayes (1797–1875), a private citizen working at the *Jardin des Plantes* who later became Professor of Conchology at the *Muséum* there. Deshayes' comprehensive study of the shells of the Paris region, which ultimately covered more than 1000 species (1824–1837), was an active project when Lyell came by in 1829 on his way back from Etna. The subtleties of Deshayes' developmental classification went far beyond Lyell's simple criteria, but Deshayes' threefold division happened to match Lyell's Eocene, Miocene, and Pliocene, and that fostered their continued collaboration.

The third volume of the *Principles* (Lyell, 1833) makes frequent reference to Deshayes' paleontology and gives long lists of Tertiary fossils, arranged in tables by age and locality, all credited to him. The question of what was and was not "a species still living" could not be resolved simply and endemism was yet to be recognized as a major problem. Subsequent editions of the *Principles*, followed in time by Lyell's hugely successful textbook, the *Elements of Geology* (Lyell, 1838), give good insight into the early difficulties of Tertiary biostratigraphy, together with a great review of all the other new developments in geology.

Stratigraphy was the number one subject for geologists all through the first half of the nineteenth century. On second look, the English layer cake turned out to be less simple than at first and much remained to be learned about the strata below the Coal Measures where Conybeare and Phillips (1822) had left off. Many people contributed to the understanding of the sequence of sedimentary rocks, but two great field geologists stand out in the geological literature of the day: Adam Sedgwick and Roderick Murchison.

From the time they had started in geology, both of them felt that Hutton's "vestige of a beginning" must be findable somewhere, and they both went to look for the oldest fossil. In the Welsh Borderland, Murchison found a continuous sequence of conformable sediments below the Old Red Sandstone, with distinctive new fossils. He worked on it for several seasons and named it *Silurian*, after an ancient tribe which had lived in that region in Roman times. Except for a few stems in the uppermost part, the Silurian contained no land plants, and the importance of that discovery was not lost on Murchison. Coal was the most important economic mineral then, and being searched for in all "Transition" strata. Now it was clear that there was no point in looking for coal where Silurian fossils were found.

Sedgwick, in the meantime, was working farther west in Wales and finding still different fossils, older than Silurian. He called them *Cambrian*, and in friendly collaboration he and Murchison examined the relationship of Cambrian and Silurian strata in the field. Also together they studied the curious sequence in Devonshire (Rudwick, 1979), where De la Beche (1839) had reported Carboniferous plant fossils in the graywacke that was then thought to be much older than the Coal Measures. The controversy that ensued confirmed that the rocks with the plants were indeed Carboniferous, downwarped in the core of a syncline of older rocks, but the older strata were not the Old Red Sandstone as would have been expected, but a sequence of slaty marine sediments with another new and characteristic fauna (Sedgwick and Murchison, 1840). They named it *Devonian*, and Murchison was able to prove, in an expedition to Germany and Russia, that this was the marine equivalent of the Old Red Sandstone (Sedgwick and Murchison, 1842; Rudwick, 1979).

As he grew older and, coincidentally, richer, Murchison also became more difficult. Sedgwick liked to confirm his field results carefully before he published them, but Murchison thought he was procrastinating. Then he decided that he saw an overlap of his Silurian with Sedgwick's Cambrian, and having made a few biostratigraphic mistakes in the field, he began insisting first that some, then that most, and finally that all of the Cambrian was part of his Silurian System. When he became Director of the Geological Survey, succeeding his old Devon adversary De la Beche in 1855, he gained a tactical advantage over Sedgwick and was able to assure the perpetuation of his claims in official maps and reports. What had started out as a friendly scientific dispute between working colleagues then turned into a protracted battle between two grand old men which did not end even when they died, a little more than a year apart; it was kept alive by their friends and disciples for several years more.

The controversy abated only when Charles Lapworth (1842–1920), in a brilliantly convincing paper (Lapworth, 1879), proposed a division of the lower Paleozoic strata into three systems: the Cambrian, the *Ordovician*, and the Silurian, naming the new system after the Ordovices, another tribe that once lived in North Wales, north of the Silures.

In 1841 Murchison went to the Ural Mountains looking for undisturbed sections he could compare with the British sequence. There he found a great accumulation of fossiliferous sedimentary rocks above the Carboniferous which appeared to have no equivalent in Britain. He called it *Permian*, after the region around the town of Perm (Murchison, Verneuil, and Keyserling, 1845).

The publications of Sedgwick and Murchison, and especially Murchison's elegant book *The Silurian System* (1839), had a great influence home and abroad. Near Prague, a French *polytechnicien*, Joachim Barrande (1799–1883) was surveying a projected railway to Pilsen (now Plzeň) in 1833, and found some interesting trilobites along the route. They aroused his curiosity and when a copy of Murchison's book reached him some time later, he became addicted to fossils (Krejčí, 1884). For the rest of his life he collected and systematically described the fossils of the Paleozoic rocks of Bohemia and published them in 30 quarto volumes (1852–1902). Following Murchison, he continued to include them all in the Silurian, even after he realized that the Bohemian section reaches from the Devonian well down into the Cambrian.

On the other side of the world in Albany, New York, James Hall, Jr. (1811–1898) burst upon the geologic scene in 1837 and began a lifetime of research on the Paleozoic fossils of New York. He also read *The Silurian System*, also published a long series of great quarto volumes, but unlike Barrande chose not to use the term *Silurian*, at least in the beginning (Clarke, 1923, pp. 80, 98). He and his colleagues decided to set up a locally derived nomenclature of their own, independent of European models.

It was another case of the right stimulus reaching the right people in the right places at the right time. Separated by 90 degrees of longitude, Barrande and Hall each happened to find himself in the middle of an exceptionally favorable area hardly touched by a paleontologist. Between them they demonstrated the enormous complexity and fantastic variety of Paleozoic marine life.

CHAPTER NINE

The French Reaction

FROM CUVIER TO AGASSIZ

By a decree of the revolutionary National Convention in 1793, the *Jardin du Roi* became the *Jardin des Plantes*, and the *Cabinet du Roi*, at least the part that Buffon had built up so assiduously, consisting of "natural curiosities," became the *Muséum d'Histoire Naturelle*. The head of Louis XVI, the former owner of it all, had rolled, the Terror had run its short but bloody course, Revolutionary armies gained the upper hand against a multitude of insurrections and foreign invasions, and the Revolution began to run out of steam. The tremendous political convulsion finally came to an end with the spectacle of Napoleon proclaiming himself emperor in 1804, but the geological work in the *Jardin* had continued through all those upheavals comparatively undisturbed.

Lavoisier was guillotined because he had once been a tax collector, and Malesherbes because he had the audacity to act as the king's defense attorney in the trial for his life. Haüy spent some time in jail, working on his crystals without interruption, it is said, but was released, returned to the *Muséum*, and continued working on his crystals. Almost all the *encyclopédistes* were dead by now, and their progressive ideas seemed to have great trouble surviving them. A new orthodoxy was about in France, but in its outlook on geology, the new was not as new as the old had been.

The dominant figure in the *Jardin* was Georges Cuvier (1769–1832), a Lutheran from Montbéliard, a French-speaking Protestant enclave in the middle of France, governed until 1793 by the Dukes of Württemberg. He was an alumnus of the *Höhe Karlsschule* in Stuttgart, a conservative bred in the Revolution, and a secretive genius full of contradictions. He had a brilliant analytical mind, a memory like a computer, and the political savvy of a Macchiavelli and a Richelieu rolled into one. Cuvier served Napoleon, cultivated Louis XVIII, idolized Charles X, and was mourned by Louis Phillipe. He had an anxious day or two in July 1830, when a popular uprising forced Charles to leave town in a hurry (and with him his grandson's tutor, one Joachim Barrande), but otherwise he was always in great favor at the Court, and never short of public funds to maintain his widespread research establishment.

Cuvier had first appeared in the *Jardin* in the spring of 1795, had impressed them all with his ability, and before the year was out was made Professor of Zoology at the University and given a parallel position at the *Muséum*. That position carried with it an apartment in the *Jardin*, near the foot of rue Jussieu, and that is where he lived for the rest of his life.

He was primarily a zoologist. At the *Muséum* he assembled a collection for comparative anatomy that still remains one of the world's greatest. The parallel collection of fossil vertebrates is magnificent and testifies to the magnitude of his work as a paleontologist. Cuvier's ability to identify and reconstruct whole animals from a few fragmentary bones became legend and his permanent contribution to geology rests almost entirely on his reliable recognition and classification of vertebrate fossils.

One of the denizens of the *Jardin* who recognized Cuvier's ability and who had helped him obtain the appointment there was Jean Baptiste Pierre Antoine de Monet de Lamarck (1744–1829), soldier turned botanist, protégé of Buffon, militant free thinker, and imaginative theorist. He had held various positions in the *Jardin du Roi* from 1788 and became Professor of Zoology at the newly organized *Muséum* in 1793.

Lamarck could hardly have been aware of Hutton's work when he gradually became convinced of the dynamic nature of the constantly-changing Earth and of the humanly-imperceptible slowness of geological processes. Major changes do take place, but they require a very long time. He did most of his fieldwork in the vicinity of Paris, where even the oldest rocks are still geologically young, but the conceptual scheme he developed comes remarkably close to Huttonian uniformitarianism. In his *Hydrogéologie* (Lamarck, 1802*a*, 1964) he shows how strongly Buffon had influenced his thinking about the oceans, and how his own ideas about the structure of the Earth developed even as he defended and wrote about them (Carozzi, 1964). He writes about the transitory nature of continental shorelines and about uplift and mountain building in a manner that makes sense today but obviously seemed bizarre to his contemporary forcing him to publish the *Hydrogéologie* at his own expense.

He was no great stratigrapher, but observed that the fossils in older rocks tend to represent lower orders of the taxonomical scale. In contrast with Cuvier's developing ideas about extinction of species, Lamarck reached the view that the division of plants and animals into defined species was artificial because all nature was dynamic and species were changing in time, like everything else. He was aware of the effect of climate on the growth of plants and thought it possible that different species might develop from the same seed in different environments, but that was only a minor aspect of the evolutionary scheme which was germinating in his mind. He considered the genetic transmission of adaptive traits, but that famous "Lamarckian" giraffe, stretching to browse on higher branches and then having offspring with longer necks, was a much later idea broadcast by his interpreters long after his death.

The purpose of the selective resurrection of Lamarck's ideas by the neo-Lamarckians half a century later was to oppose Darwin and to make evolution compatible with the teleological argument of design and harmony in nature. Lamarck himself never used the word *evolution* and his own expression of evolutionary systems was neither clear nor tidy (Lamarck, 1802*b*, 1815–1822). It produced little response in his own time apart from transforming his early friendship with Cuvier into enduring hostility. The thought that living species may have developed from fossil ones was totally unacceptable to Georges Cuvier.

Lamarck's ideas on evolution were carried even further by his junior collaborator Étienne Geoffroy Saint-Hilaire (1772–1844), a protégé of Haüy, his romantic defender when he was imprisoned, a student of Daubenton, and Cuvier's close friend during his first years at the *Jardin*. In 1798 Cuvier declined but Saint-Hilaire accepted the post of naturalist with Napoleon's army in Egypt. He stayed three years and brought back, among many other things, a collection of mummified animals found in pharaonic graves. He was able to show that the creatures of 3000 years ago were taxonomically identical with modern representatives of those same species, and to Cuvier that was proof that species were fixed, whereas Lamarck was convinced that 3000 years was too little time to prove anything.

Chain-of-being philosophies seeking to observe unity in nature by arranging it all on a scale of progressively "higher" steps, from simple minerals on the bottom to man at the top, were often discussed in Germany when Cuvier was a student. However, Cuvier's conservative outlook and his eagerness to comply with Napoleon's new orthodoxy in religion led him to reject any such thoughts as blasphemous and contrary to the Word of God. Cuvier was convinced that species were fixed, immutable, and independent. To him, the only connection between fish and mammals was that they both had come from the hands of the same Creator. That position was clearly incompatible with the expressed views of Lamarck and Saint-Hilaire, and the former friends became bitter enemies.

Cuvier's sneaky and vicious attacks on Saint-Hilaire remain one of the classics of personal warfare in science. In the eyes of their contemporaries, Cuvier came out the overwhelming winner, but it is now obvious that he was wrong. It remained for Darwin to show much later that Lamarck and Saint-Hilaire were on the right track and that Cuvier's own paleontological results are entirely consistent with the theories he so vehemently denounced.

The fourth star player on that stage was Alexandre Brongniart (1770–1847), who had started as assistant to his chemist uncle Antoine-Louis Brongniart (1742–1804) at the *Jardin*, followed a career as a mining engineer in the government, and in 1822 was appointed Haüy's successor as Professor of Mineralogy at the *Muséum*. When Cuvier began to reconstruct and describe his fossil menagerie from the Paris basin, Brongniart became interested in the stratigraphic relationships of these

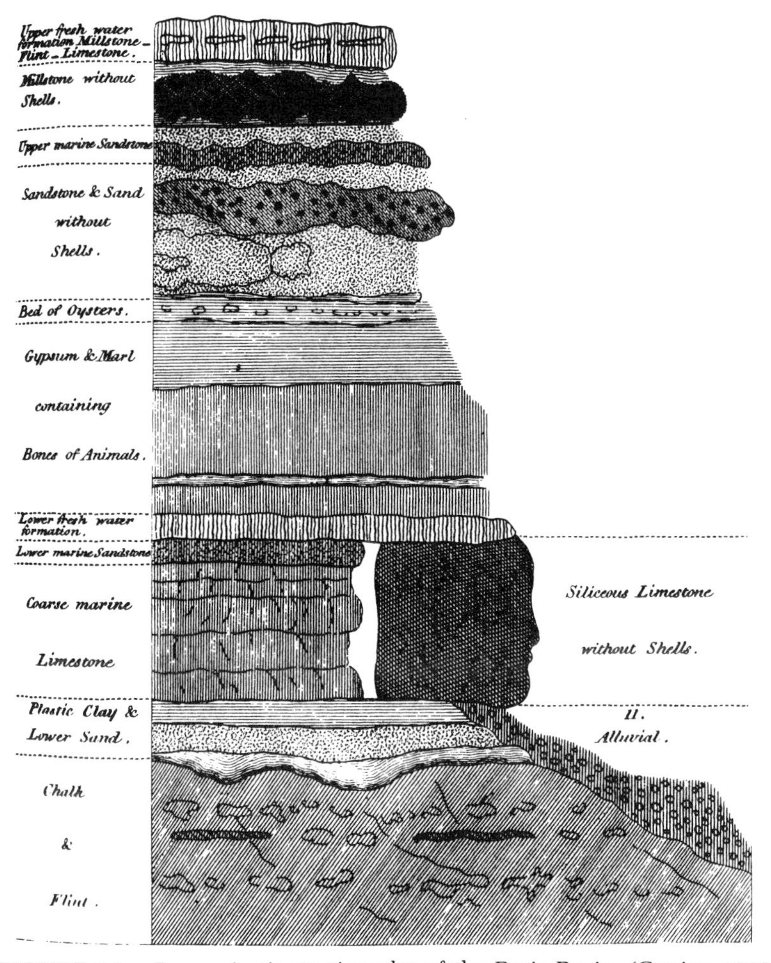

FIGURE 9.1. Brongniart's stratigraphy of the Paris Basin. (Cuvier, 1818)

beasts and the two of them collaborated in mapping the region and working out the sequence of strata (Cuvier and Brongniart, 1808–1811). Cuvier freely acknowledged that the work was mostly Brongniart's, but it was to be expected that Cuvier's name would come first.

Brongniart's biostratigraphy of the Paris basin made it obvious that Werner's classification of sedimentary rocks into Transition (dipping) and Secondary or Floetz (flat-lying) was fundamentally invalid. The soft chalk of the Paris basin was very nearly flat, but the fossils in it were equivalent to the fossils found in some hard and steeply-dipping limestones in the Alps. Lithology and dip offered no help with the sequence of strata, but fossil assemblages did. When Brongniart surveyed the Paris basin beginning about 1804, he was almost certainly unaware that William Smith in England had reached the same conclusion some years before but was yet to publish it.

The continuing research produced additional evidence that the stratigraphically lower (and thus older) animal assemblages were also lower on the taxonomic scale and included many forms that are now obviously extinct, even on the family level. How could these extinctions be explained? Most of the strata of the Paris basin (Cretaceous and younger) had been deposited in the sea, judging by the fossils, but two formations contained freshwater clams and the lower of the two had continental vertebrates. For some mysterious reason, the sea had drained from the land twice since the deposition of the chalk.

The demonstrated return of the sea had to have had a catastrophic effect on the animals that roamed the land, and to Cuvier, this would explain the extinctions. "These repeated irruptions and retreats of the sea have been neither slow nor gradual; most of the catastrophes which have occasioned them have been sudden," (Cuvier, 1813). One proof of the suddenness of such events was the presence of unputrefied carcasses of large extinct animals in the northern ice. He presented these views in the introduction to his great study of fossil quadrupeds (Cuvier, 1812*a*), which he also published separately as the *Discours sur les Révolutions du Globe* (Cuvier, 1812*b*).

Like other genuinely religious geologists, Cuvier did not care for theorizing and generalization. He refers to his *Discours* as a "Theory of the Earth," but it is actually a sober discussion of Brongniart's geology, strongly influenced by Buffon, and not warped to conform with Genesis even if the manner of presentation appears calculated to permit others to make such connections. It dwells on the importance of fossils in the study of the history of the Earth, defends the fixity of species in stratigraphic terms, and lets the readers make their own conclusions. The book was a huge success, was translated into a multitude of languages, and appeared in many editions.

Robert Jameson translated it into English (Cuvier, 1813) and added a lot of notes that try to make Cuvier's statements appear in Wernerian light. He correlates Cuvier's last extinction with the Noachian Deluge and implies many other ideas that go far beyond the author's original intentions, but that is the form in which Cuvier was being presented to the English-speaking public. In New York, Samuel Latham Mitchill, MD (1764–1818), professor at Columbia College, author of the first American syllabus for a course in geology (Mitchell, 1792), a prime mover of geological affairs in New York (and later U.S. Senator) brought out another edition of Jameson's translation (Cuvier, 1818). He added still another appendix with much geological information and conjecture but without theological overtones. That is how Georges Cuvier, who established vertebrate paleontology as a science, was made to appear as a reactionary to British and American geologists.

Cuvier was 61 when he met Jean Louis Rodolphe Agassiz (1807–1873), the son of a Protestant clergyman from French Switzerland, aged 23, recent MD from Munich, and, more importantly to Cuvier, the author of a very good dissertation on Brazilian fishes. Young Agassiz became not only a confirmed disciple in zoological taxonomy but also the most persistent and uncompromising defender of Cuvierian ideas on extinction and the fixity of species.

Unfortunately, the master soon died without having provided for the disciple, and Agassiz had to return to the provinces. He became a professor at the new College of Neuchâtel in 1832, and there distinguished himself both as a scientist and as an administrator. He produced a monumental treatise on fossil fishes (Agassiz, 1833–1844) and a study of glaciers which presented to the geological world the startling news that those erratic boulders, jumbled sediments, and U-shaped valleys that had been observed in so many places far from any present-day glaciers were produced by glaciers that once covered the continents. (For a lucid history read Davies, 1969).

The idea had originated about 1820 with Ignatz Venetz (1788–1859), a Swiss civil engineer, who convinced Johann de Charpentier (1786–1855), a Swiss mining superintendent and graduate of the Freiberg Academy. They could not convince anyone else until they took Agassiz into the field in the summer of 1876, and made him a militant convert. With overwhelming enthusiasm Agassiz developed glacial geology and demonstrated his results in an elegant publication *Untersuchungen über die Gletscher* (Agassiz, 1841) that brought him international fame. In his own mind the Great Ice Age he postulated would also serve admirably to explain Cuvierian extinctions.

On visits to England, Agassiz had found a wide area of mutual understanding with Cuvier's British admirer, William Buckland, but when Buckland came to Switzerland in 1838, it took him some time in the field before he could believe that glaciers really did all those tremendous things Agassiz said they did. When two years later Agassiz visited Buckland, they toured Scotland together and found evidence of glacial action almost everywhere they looked. That same summer Buckland demonstrated the same features to the skeptical Lyell, and thought he had convinced him.

On the strength of these accomplishments, Agassiz was invited to lecture at the Lowell Institute in Boston in 1846, and then accepted a professorship at Harvard. He prospered in the Yankee intellectual climate, where seeking conformation of geology with Genesis was a valid pursuit long after such efforts had been dropped by serious scientists elsewhere. His Cuvierian attitude developed into an unrelenting (Agassiz and Agassiz, 1868) and tragically disillusioning (Gould, 1979) campaign to contradict Darwinian evolution. Like his great master, Agassiz made very important contributions to geology, especially glacial geology, but he is remembered today for his failure to recognize the overwhelming significance of evolution.

CHAPTER TEN

Provincial Beginnings in America

TRAVELERS AND NATURALISTS

Passing mentions of American geology are fairly common in the writings of travelers who visited the new land and the few naturalists who lived there, but the total amount of geological intelligence that could be gleaned from all of them is very small. None of the writers was a geologist even secondarily. Noteworthy among them are the English astronomer Thomas Harriot (1560–1621); the explorer-naturalist John Banister (1650–1692); the planter and historian Robert Beverley (1673–1722); the botanists and nurserymen John Bartram (1699–1777) and his son William (1739–1823), who also became known as a writer; the cartographer Lewis Evans (1700–1756); the founder of almost everything in Philadelphia, Benjamin Franklin (1706–1790); the Harvard astronomer and Fellow of the Royal Society John Winthrop (1714–1779); the Finnish botanist and traveler Peter Kalm (1716–1779); and the British colonial administrator Thomas Pownall (1722–1805).

Many explorers and missionaries traveled the Great Lakes region (Foster and Whitney, 1850, pp. 5–13; Wells, 1963), and important among them is Father Louis Hennepin (1640–?1701), not for his exaggerated claims of discovery, but because he gave the first true illustration of Niagara Falls (Hennepin, 1697). These early reports and relations were searched for "mineralogical" data by Jean Étienne Guettard, who summarized them on a map (Guettard, 1752*a*). Technically, that was the first geological map of North America (Cailleux, 1979), but it gives only isolated localities of minerals and rocks without geological interpretation.

The New England earthquakes of 1727 and 1755 produced a flurry of

FIGURE 10.1. Niagara Falls. (Hennepin, 1697)

sermons, many of which were published. Their content is almost entirely theological and shows, more than anything else, how very far removed New England clergymen were from any thoughtful study of nature. The only attempt at scientific analysis of earthquakes came from John Winthrop (1755) at Harvard.

Benjamin Franklin had opinions on geology. He knew about fossil shells found high in the Appalachians and columnar basalt near New Haven. In 1788, in a letter to Giraud Soulavie, Franklin entered into some wild speculation on the possibility of the interior of the Earth being filled with a dense liquid; he was undecided about the nature of the liquid, but thought it might be compressed air. Franklin was always concerned and involved with scientific activity, whether he was in Paris, London, or Philadelphia. He was a good friend of the Bartrams, Evans, Kalm, and Pownall, and a central figure in the American Philosophical Society.

Many minor geological articles appeared in the growing American scientific literature after the Revolution (Hazen and Hazen, 1980), but the geologic picture of America one could derive from a synthesis of all these accounts would hardly be a picture at all. There was no illustrated description of any fossil, not a glimmer of stratigraphy, and very little mineralogy. When the eighteenth century came to an end, even the eastern parts of North America would still have been well described as a geologic *terra incognita*.

The only person to have recorded comprehensive observations of the geology of the United States by that time was a young Hessian surgeon, Johann David Schöpf, MD (1752–1800). He arrived in New York in 1777, spent six years "confined to the narrow compass of sundry British garrisons along the coast" (Schöpf, 1788; 1911, p. 11), and then, instead of returning home with his regiment, set out on a year's journey of naturalistic exploration which took him overland through New Jersey, Pennsylvania, Maryland, Virginia, and the Carolinas to Charleston, and then by sea to St. Augustine and the Bahamas. "To be candid, the motive of my journey was curiosity" is the way he explains it (Schöpf, 1911, p. 4).

It was the kind of curiosity that generates scientific exploration. Schöpf had taken courses in botany, zoology, and "mineralogy" at the University of Erlangen, was an uncommonly keen and analytical field observer, and knew how to present his results in a logical manner. He observed the landforms, recorded where what kind of rock was found, closely examined its field relationships and its mineral composition, and then made the best possible effort to explain what he had seen. He understood that granite could not precipitate from the water of the ocean, and considers it part of the "Primitive" crystalline basement.

Schöpf does not cite Lehmann or Arduino, but their ideas seem to have been part of the teaching at Erlangen. He points out how the sequence of formations in the Appalachians could be shown "in different colors on the general map" (Schöpf, 1787*a*, 1972, p. 80) but realized that it would take more traverses than he could make in the time he had available. These were advanced ideas in 1787, and they indicate that the geological curriculum at the then small and relatively unimportant University of Erlangen was very strong and its students rigorously trained.

In addition to the account of the geology of the United States from Rhode Island to the Carolinas, the young doctor also brought out a two-volume book of his travels (Schöpf, 1788, 1911). It is arranged chronologically and also contains some geological information but is valuable mainly for the warm, witty, and perceptive manner in which it paints the life of post-revolutionary America. Having been one of the enemy, and a Hessian besides, he had to move with some caution, but he managed to collect an astonishing amount of information. He published an American *materia medica* (Schöpf, 1787*b*), an illustrated book on turtles (unfortunately unfinished), a long article on American fishes, and many shorter articles (Spieker, in Schöpf, 1972).

Judging by the good representation of these works in old European libraries, it would appear that they were widely read in German-speaking countries. German immigration to the United States was increasing and German interest in the new republic was very high at the time. Many Germans already lived in the United States, especially in Pennsylvania, but German was not a popular foreign langauge among American intellectuals. Schöpf's *materia medica* was in Latin, and did receive some attention here, but his geological work went unnoticed until 1842, and then was mentioned only in passing as a curiosity (Association, 1843, p. 69).

The precarious fiscal state of the new republic made reconstruction of the wartime damage slow. Things were just beginning to improve a dozen years after Schöpf's trip when Constantin-François Chasseboeuf, later Comte de Volney (1757–1820), arrived in the United States. He was a French gentleman-traveler, a former friend of the Revolution who had spent 10 months in jail during the Terror and, thoroughly disillusioned, emigrated to the United States in 1795. He traveled for three years and made a long trip far into the interior to Detroit, Cincinnati, Louisville, and Vincennes (on the Wabash), mostly by riverboat, seeking out French settlers and making it a point to meet important people. Being not the most tactful of Frenchmen, he also made some important enemies (notably President Adams) and feeling unwelcome, returned to France. There he published his two-volume work on the "climate and soil" of the United States (Volney, 1803, 1804, 1968).

Volney was neither geologist nor naturalist, but he was acquainted with publications in those fields in French and English and was known in scientific circles. For example, he did not hesitate to send Lamarck the fossils he collected in America for identification. Evidently he never heard of Johann David Schöpf and had neither his penetrating eye nor his open mind, but he did publish a broad geological description and the first real colored geological map of the eastern United States. Following Samuel Mitchell a few years before, he divides the country into geologic "regions" of granitic rocks, sandstones, calcareous rocks, marine sands, and river deposits, and outlines their geographic extent in his text.

Volney's map shows two geological features: a "calcareous vein" stretching from Washington to present-day Winston-Salem, and a "Banc d'Isinglass" (micaceous bed, translated in the English edition of 1804 as "Ridge of Talcky Granite") from Perth Amboy through Trenton, Philadelphia, Baltimore, Washington, and Richmond, to Halifax in North Carolina. The map has no guide lines or explanation of colors, but at the end of the book (Volney, 1803) a curious "advice to the reader" says that the map is colored and explains the meaning of the colors. Just the same, the book was sold to the general public with the map uncolored, and only a few special large-paper copies have been found with the map in color (Wells, 1959; White, 1977).

One of these special copies Volney gave to his friend William Maclure (1763–1840) and it became an inspiration to him. Maclure had made his fortune in business in London, became interested in geology, and turned scientific organizer and promoter of educational schemes. Like many other activists, he was a person of preconceived notions and excessively motivated by immediate short-term considerations. He was concerned with transport, mineral resources, and the fertility of soils, and set out to survey the United States to evaluate their territory for agricultural and commercial development. Maclure believed that these goals would be served by a gross geological reconnaissance outlining large areas of equivalent rock types on a grand scale (about 75 miles to the inch in the first edition of his map). "The geologist must endeavor to seize the great and prominent outlines of nature; he should acquaint himself with her general laws, rather than study her accidental deviations, or magnify the number and extent of the supposed exceptions, which must frequently cease to be so when judiciously examined" (Maclure, 1809*a*, pp. 427–428).

Maclure's short article in the *Transactions* of the American Philosophical Society (Maclure, 1809*b*) shows signs of having been written in haste and the map is crude by contemporary standards of map making. It was more an exhibit than an aid, but it was to become the first frequently reprinted and widely distributed geological map of the eastern United States (Maclure 1809*b*, 1817, 1818*a*; Cleveland, 1816, 1822; see White, 1977). In that light it was widely influential as a symbolic point of departure for the geological exploration of the American continent.

As a geologist, Maclure was a Wernerian by way of Jameson. He subdivides the rock formations of the United States according to the Wernerian system and describes them in Wernerian terms, but makes a point of avoiding "entering into any investigation of the origin or first formation of the various substances." His geognostic classification is taken from Jameson (1808, pp. 99–100), word for word, leaving out only "Class V., Volcanic Rocks." He was surprised to find so little volcanic rock in the United States. His geological information comes partly from direct observation, but also from Mitchill (1798, 1802), Volney (1803, 1804), and many other sources, some reliable and some not. He mentions many mines and mineral localities, but from his own words it is difficult to decide whether he is reporting his own observations or hearsay, and he acknowledges none of his sources. The only geologist's name cited in the first paper (Maclure, 1809*a*) is Werner's.

Maclure owes a debt to Volney, yet their approaches are quite different. Volney reports, in gross outlines, what kinds of rock he saw and estimates their extent beyond his own areas of observation by what he read or heard, but Maclure is out to present a system. What Volney's map shows as "micaceous bed" and "calcareous vein," roughly corresponding to what is now called the Wissahickon Schist and the several calcareous formations to the west of it, Maclure lumps under "Transition," and what Volney colors as "limestone" and "sandstone" (on the rare large-paper version of his map), Maclure shows as "Secondary." His sincerity in wishing to avoid "any investigation of the origin" need not be doubted, but the genetic implications of the Wernerian terms were fully established then and he could hardly have hoped to make them vanish.

"The foregoing observations," writes Maclure (1809*a*), "are the results of many former excursions in the United States, and the knowledge lately acquired, by crossing the dividing line of the principal formations in twenty-five or thirty different places, from the Hudson to Flint River" (in Georgia). Eight years later he repeats the same phrase exactly (Maclure, 1817, p. 58; 1818*a*, p. 41), thus depriving it of some credibility. The 1817 article repeats the 1809 version almost verbatim, and it is made longer only by the addition of a long discussion of the soils found in the United States and by many general comments and explanations, with very little new information on the geology.

Maclure had regular access to the *Transactions* of the Geological Society of London (he was elected Foreign Member in 1823) and must have been well informed of the rapid development of geology in Europe at the time, but his own thinking changed only slightly in the face of all that new information. He continued to proclaim faithfully that "the system of Werner is still the best and the most comprehensive that has yet been formed" (Maclure, 1817, p. iii), but finally realized that he could not go on deliberately ignoring "the origin or first formation of the various substances."

He published a detailed summary of his views on the formation of rocks, still in Wernerian outlines (Maclure, 1818*b*,) but modified by the concession that basalt is volcanic and that granite "has more resembance to some of the feldspathic lavas, than it has to any rock known to be of Neptunian origin. It likewise approaches the volcanic, in relative situation, without any regular stratification. Yet the resemblance does not appear sufficiently strong to amount to direct analogy, and we must therefore remain in doubt as to the nature of its origin" (Maclure, 1818*b*, p. 340).

In terms of Maclure's great influence on American geology, this turns out to have been a minor point. More significantly, and unlike his French and English contemporaries, Maclure did not see the importance of fossils in stratigraphy. Werner's system asserted that rocks with inclined bedding ("Transition") were older and basically different from rocks that lie flat ("Secondary"). That approach made it impossible for him to even begin to understand the relationship of the formations in the folded Appalachians to the flat-lying strata further west.

Near Lewiston, in the Niagara Gorge, Maclure reports a reddish sandstone formation (now known to be a member of the Silurian Albion Sandstone) and understandably mistakes it for an equivalent of the Old Red Sandstone of Britain, where it was known that coal-bearing strata can be found above the Old Red. Maclure knew the value of coal, but he did not appreciate how important it was becoming. He viewed the United States as an agricultural country and ignored the increasing signs of industrial development. He opposed the Erie Canal project because he thought "that it is probable the whole surplus produce that would pass through it would not pay one per cent on the sum expended, in making it" (Maclure, 1817, p. 98; 1818*a*, p. 70). At the same time he thought that his "Old Red" sandstone "is the foundation of all this horizontal formation" in western New York (Maclure, 1817, p. 57; 1818*a*, p. 40), and thus misled the promoters of the region into thinking that coal may be found in the flat-lying formations that cover that large part of the state.

GEOLOGY AT AMERICAN COLLEGES

Compared with their British counterparts, American colleges did not amount to much in the beginning of the nineteenth century, but they made up in determination what they lacked in resources—both academic and financial. Their dynamic outlook, coupled with their usually precarious fiscal position, combined to make them open to suggestions from the community, and their response to the growing demand for training in the arts and sciences was direct: They did their best to provide such training.

Philadelphia was then the focal point of American business, politics, and science. The American Philosophical Society had been founded there in 1743, merged with the American Society for Promoting Useful Knowledge in 1769, and published the first volume of its *Transactions* in 1771. Benjamin Franklin, David Rittenhouse, and Thomas Jefferson were the first three presidents of the merged societies; all three of them much interested in promoting useful knowledge.

A few blocks away, the University of Pennsylvania was developing the medical school, striving, in a modest way, to approach its model, the University of Edinburgh. It was there that James Woodhouse (1770–1809) began to give occasional lectures in chemistry, mineralogy, and geology, about the turn of the century. Jefferson's friend, "that talented madcap" Thomas Cooper (1759–1839), taught applied chemistry and mineralogy from 1815 to 1819 before he moved on to South Carolina College. Both Woodhouse and Cooper were exceptionally clear thinkers. It was Cooper who analyzed the density of the minerals that make up the Earth's crust, considered the overall density of the Earth as determined by Maskelyne and Cavendish, and concluded that the Earth must have a metallic core, probably consisting of iron and nickel (Cooper, 1813).

A Course of Lectures at the University of Pennsylvania, 1817

About the middle of October, Judge Cooper proposes to commence his Geological and Mineralogical Lectures in the University of Pennsylvania. They will consist of the following parts:

1. Introductory Lecture.
2. On the Globe of the Earth: on the general properties of Mineral Substances; specific gravity, hardness, fracture, chrystallization, colour, &c.
3. On the rocks called *Primitive*, and their component parts.
4. On the Substances found in Primitive rocks.
5. On the rocks termed *Transition*, and their component parts.
6. On the Substances found in Transition rocks.
7. On the rocks termed Secondary.
8. On the Substances found in Secondary rocks.
9. On Alluvial Formations.
10. On Basins. The great Basin of the Mississippi. The Basin at Richmond, Virginia. The Paris Basin. The London Basin. The Isle of Wight Basin.
11. On Volcanic Formations. On Floetz Trap.
12. On Organic Remains.

This course of Mineralogy, which, as the reader will see, is very different in its outline from any hitherto attempted, will be illustrated by appropriate specimens, Judge Cooper's cabinet, being now, the best adapted for the purpose, of any in the United States, colonel Gibbs's excepted. To which gentleman, and Mr. Maclure, Judge Cooper expresses his obligations for the kind assistance they have afforded him in this respect.

This collection of between three and four thousand specimens, consists of his own collection; of the late Rev. Mr. Melsheimer's, and of M. Godon's.

It is expected the Course will occupy between two and three months, at three Lectures a week. Tickets 15 dollars.

(Thomas Cooper, *The Analectic Magazine*, Vol. 10, October 1817, p. 352)

In New York, Samuel Lathman Mitchill was teaching chemistry with a lot of geology and mineralogy at Columbia College from 1792 to 1801, and at the College of Physicians and Surgeons of the University of the State of New York from 1807 to 1826, even while he was serving in the State Assembly and in Congress. His geological publications are numerous and diverse, but probably reflect much less than the full scope of his far-ranging interest. He was very active on the New York scientific scene and involved himself in many beneficial endeavors.

NOVEMBER, 1814.

FACULTY

OF THE

Academical Institution

OF

Yale College.

REV. TIMOTHY DWIGHT, S. T. D. LL. D.
President, and Professor of Divinity.

JEREMIAH DAY, A. M. *Prof. of Math. and Nat. Philosophy.*
BENJAMIN SILLIMAN, A. M. *Prof. of Chem. and Mineralogy.*
JAMES L. KINGSLEY, A. M. *Prof. of Lang. and Eccl. History.*

TUTORS.

ARÆTIUS B. HULL, A. M.
SAMUEL J. HITCHCOCK, A. M.
JOHN LANGDON, A. M.
JOSIAH W. GIBBS, A. M.
RALPH EMERSON, A. B.
WILLIAM DANIELSON, A. B.

FACULTY

OF THE

Medical Institution

OF

Yale College.

ÆNEAS MUNSON, M. D. *Professor of Materia Medica and Botany.*
NATHAN SMITH, M. D. C. S. M. S. Lond. *Prof. of the Theory and Practice of Physic, Surgery, and Obstetricks.*
ELI IVES, M. D. Adjunct *Prof. of Materia Medica and Botany.*
BENJAMIN SILLIMAN, A. M. *Prof. of Chemistry and Pharmacy.*
JONATHAN KNIGHT, A. M. *Prof. of Anat. and Physiology.*

Summary.

Resident Graduates,		16
Seniors,	74	
Juniors,	66	
Sophomores,	65	
Freshmen,	72	
Academical Students,		277
Medical Students,		57
Total,		**350**

FIGURE 10.2. Yale in 1814.

In a move that turned out far better than anyone might have hoped, Yale appointed a young lawyer, Benjamin Silliman (1779–1864), as Professor of Chemistry and Natural History in 1802, ignoring the conspicuous fact that he lacked even the slightest conventional qualifications for the job. Silliman quickly read what few chemistry books he could find in New Haven and then rushed to Philadelphia to take a cram course with Woodhouse. In 1805 he went to Europe to buy apparatus and assemble a library, spent a term at the University of Edinburgh, and met Hall and Murray. Then he returned to New Haven and methodically went about making Yale the foremost scientific school in America, with particularly strong credentials in geology.

In 1818 Silliman established the *American Journal of Science* and through it he soon became the world's best-known American geologist. He edited three American editions of Bakewell's *Introduction to Geology* in 1829, 1833, 1839, and in 1836 he corresponded with Lyell about the possibility of an American edition of the *Principles*. Those arrangements failed when an enterprising Philadelphia publisher brought out the first American edition in 1837, simply pirated from the fifth edition, which had just appeared in London. There was no copyright agreement between England and the United States at the time, and so there was nothing either Lyell or Silliman could do about that.

Silliman had three great students in geology: Amos Eaton (1776–1842), Edward Hitchcock (1793–1864), and James Dwight Dana (1813–1895). All three became influential teachers and writers and had a profound effect on the development of American geology. Eaton established his own school at Rensselaer (in Troy, N.Y.) and Hitchock at Amherst (Massachusetts), and they were instrumental in setting up the geological surveys of Massachusetts and New York. Dana made his name with the U.S. Exploring Expedition (Wilkes Expedition), became Silliman's son-in-law and heir apparent, and is now remembered as the great compiler of systematic mineralogy.

Amos Eaton's career was anything but ordinary. He began as a country lawyer. surveyor, and land agent in Catskill, New York, became ensnared in a vicious land dispute, was charged with forgery, and sentenced to life in prison. The extent of his guilt, if any, has always been in doubt (McAllister, 1941), but the magnitude of the sentence, inconceivable as it might seem today, was not unusual at the time. He was sent to Greenwich jail where, as it happened, the man in charge was William Torrey, veteran of the Revolutionary War, New York City Alderman, State Prison Agent, and the father of John Torrey (1796–1873) who was to become a great botanist and professor at Princeton, but who was only 15 years old when Eaton arrived at the jail. The boy and the lawyer studied Linnaean botany together and became lifelong friends. Eaton started John Torrey on his great career, and the young man returned the favor by persuading his father to help arrange a pardon for his teacher.

Many other friends came forward, notably Dr. Mitchill, and Eaton was released in 1815. He went to Yale for a year, then to Williams College (his old alma mater), and ultimately to Troy, where he settled even though the official pardon had specifically ordered that "he depart from the state of New York and never thereafter return to the same" (McAllister, 1941, p. 152). New York political skulduggery had not come to an end, but the wind had shifted. The governor was now De Witt Clinton (1769–1828), sponsor of the Erie Canal, friend of science, and promoter of education and penal reform.

Edward Hitchcock studied at the Deerfield Academy and began teaching astronomy there. He became interested in geology through Amos Eaton, but first served as Congregationalist minister in Conway, Massachusetts, from 1821 to 1825. Then he became Professor of Chemistry and Natural History at Amherst College, took one course with Silliman, and was on his way to a distinguished geological career, which finally led to the presidency of Amherst. He had mapped some geology in the Connecticut valley while still at Deerfield, and was named state geologist of Massachusetts in 1830 and of Vermont in 1856. The report of the Massachusetts survey (Hitchcock, 1833) was the first major monograph published by a state survey and included the first geological map of an American state.

James Dwight Dana, fresh from graduation from Yale in 1833 and needing gainful employment, signed on as a teacher of mathematics to the midshipmen of the USS *Delaware*. A few years later Silliman offered his former student an assistantship at Yale. Soon afterward he joined the Wilkes Expedition around the world where his great talents in geology and zoology developed.

Chester Dewey (1784–1867) was Professor of Mathematics and Natural Philosophy and Lecturer in Chemistry at Williams College. He also taught geology and mineralogy, had an excellent collection of minerals and fossils, and published a geologic map of Berkshire County, Massachusetts (Dewey, 1819).

At Harvard, John White Webster, MD, taught chemistry, which included some geology, beginning in 1824. In 1817 he had toured Scotland with Ami Boué (1794–1881), a student of Jameson later prominent in French geology, and on his way home stopped on São Miguel (St. Michael) Island in the Azores. There he was fascinated not only with the widespread volcanic rocks, but also by Harriet F. Hinckling, the charming daughter of the American consul. He made the first geological survey of the island (Webster, 1821), and married Harriet in 1823. His best student at Harvard was Charles Thomas Jackson, MD (1805–1880), later the state geologist of Maine, New Hampshire, and Rhode Island, and U.S. geologist in Michigan. Both of them met their ends in unfortunate ways: Jackson died insane, and Webster was hanged for murder (Sullivan, 1971; Thomson, 1971).

At Bowdoin College in Maine, Parker Cleaveland (1780–1858) was appointed Professor of Mathematics and Natural Philosophy in 1805. He had graduated from Harvard at the age of 19, considered divinity and law, but decided on science and an academic career. His reputation rests on his *Elementary Treatise on Mineralogy and Geology* (Cleaveland, 1816, 1822), which followed European models in the Wernerian tradition and included reissues of Maclure's map. It is important because it presents the first broad compilation of American mineral occurrences, which he had assembled through much correspondence. That made him well-known far beyond New England, and in 1818 he became Foreign Member of the Geological Society of London, the first American so recognized by that august body.

Farther south in Philadelphia, geology was taught with chemistry and mineralogy by James Woodhouse and Thomas Cooper, as already mentioned. Another well-informed mineralogist in town at the time was Adam Seybert (1773–1825), who graduated as an MD from the University of Pennsylvania, spent several years studying in Europe, assembled an outstanding collection of minerals, served four terms in Congress, but was unable to obtain a teaching appointment (E. F. Smith, 1919, p. 27). Both Seybert and Cooper owned copies of Schöpf's *Beyträge* (1787*a*), but neither left any trace in print to show that they had ever used it.

William Hypolitus Keating (1799–1840), mineralogist and early western explorer, and Alexander Dallas Bache (1806–1867), geophysicist and later Director of the Coast Survey, also taught at the University of Pennsylvania. Henry Darwin Rogers (1808–1866) was in residence there (without salary), and at the same time directed the first geological survey of Pennsylvania. Timothy Abbot Conrad (1803–1877) and Isaac Lea (1792–1886), paleontologists and biostratigraphers, also lived and worked in Philadelphia.

Still farther south, geology was taught at the College of William and Mary in Williamsburg, where Patrick Kerr Rogers (1776–1828), emigré from Ireland and MD from the University of Pennsylvania, became Professor of Natural Philosophy and Chemistry in 1819. His four sons, James (1802–1852), William Barton (1804–1882), Henry Darwin (1808–1866), and Robert Empie (1813–1884), received their first education there and went on to fame in the sciences.

Thomas Cooper moved from Philadelphia to the University of Virginia and then to South Carolina College, where he was president from 1821 to 1834, and where Lardner Vanuxem (1792–1848) taught from 1819 to 1826. Denison Olmsted (1791–1858), another student of Silliman, taught chemistry, mineralogy, and geology at the University of North Carolina from 1819, and he has the distinction of having promoted the first state geological survey in the United States in 1823. The budget of his survey was $250 per year and that may be one reason why two years later he was back at Yale, this time as Professor of Mathematics and Natural Philosophy. He did little more geology, but wrote several successful textbooks in natural philosophy.

Great Britain was still very much the mother country in 1820, but America was going its own way. Even before westward expansion, the territory of the United States seemed enormously large to Europeans, and the population very small. Agriculture was still the dominant industry, and it was conventional wisdom to predict that the future of the country lay in that direction.

Geology was developing largely through the progressive outlook of the colleges. There was no ready supply of diligent well-to-do physician-naturalists and subsidized clergymen-geologists to do the fieldwork and gather the background data just for fun as in Britain. With few exceptions, American physicians were occupied with their practice, and the American clergy, mostly struggling in meager circumstances, did not share the liberal inclinations of their British colleagues. Mosaic chronology and Noah's Flood were taken as fact in America much longer than in Britain, and this helps explain the long survival of Wernerian views on this side of the ocean. The abstract theism of Hutton and Playfair did not fit that picture at all, and their ideas could not have been expected to make much headway against the neptunist trend. (See, for example, Cooper, 1813; Cleaveland, 1816, pp. 593–594; H. H. Hayden, 1820, pp. 254–255; Van Rensselaer, 1825, pp. 35–37).

RIVERS AND CANALS

More and more people were moving westward and inland transportation was growing rapidly. Navigation on the Ohio and the Mississippi rivers made Pittsburgh the logical starting point for western exploration; construction of the Erie Canal in 1817–1825 established Buffalo as the port of trade on the Great Lakes. There were hints of important mineral deposits in the West: lead in the southeastern part of the Missouri Territory and copper on Lake Superior, but actual exploration took a while to get started.

Wood was still a plentiful fuel, but its heat content was low and labor to cut it became increasingly expensive. Iron was still being made with charcoal, but the coalers used up the wood faster than it grew back and the distance from usable timberlands to the furnaces in many places grew longer and longer. Steam engines appeared here and there and proved remarkably effective, but they needed a lot of fuel. The obvious answer to these demands for energy was coal, just as it had been in England a little earlier, but American distances were much greater, and coal was not always found where it was needed. It took American geologists some time to learn how to locate coal, but learning it helped to establish the value of geological surveys.

The first such survey was ordered by the enterprising "Patroon"

FIGURE 10.3. Entrance of the Erie Canal into the Hudson River at Albany. (Eaton, 1824)

Stephen Van Rensselaer (1764–1839), who owned large tracts of land in the upper Hudson Valley and was one of the main promoters of the Erie Canal. Amos Eaton was engaged to do the job single-handedly, and he covered the territory from the Massachusetts line to Lake Erie in just one year, 1823. For the benefit of future students, he wrote that "by merely taking carriage to Williams College (in Massachusetts), thirty miles, and eighteen miles along the south shore of Lake Erie (to Eighteen-Mile Creek, south of Buffalo), all our rocks may be seen in place, by the traveler in canal packets and steam-boats" (Eaton, 1824, p. 13).

Unlike William Smith, his distinguished predecessor in the canal survey business, Amos Eaton did not appreciate the importance of fossils. Following Maclure, he adhered to the Wernerian notion that flat-lying rocks are Secondary, and therefore must be young. His model for stratigraphic nomenclature was *Conybeare and Phillips* (1822), but they did not cover strata below the Carboniferous. In England, at that time, not much was known about the stratigraphy below the Coal Measures, but Eaton was not aware of that. He thought that the "true Coal formation" was to be found in the flat-lying sedimentary rocks of western New York, and he did not realize that the fossils he was collecting along the Erie Canal were much older than that.

Eaton's great importance to American geology lay not so much in his surveys but in his determined drive to teach geology and popularize it. He wrote many small, simple textbooks and had them printed in Albany, where printing was inexpensive when the legislature was not in session. That made the books inexpensive and thus widely accessible.

Farmers and businessmen became aware of the value of geology through his efforts, and it was this awareness that led to the establishment of geological surveys in so many states. The lobbying of professional geologists was important in the legislatures and the public's broad interest in the study of nature was evident, but the force that kept the survey movement going was pressure from real-estate and commercial interests aroused by popular education in geology.

Eaton was instrumental in the organization of the Rensselaer School in Troy, which opened in 1824 and in time became the Rensselaer Polytechnic Institute, the oldest engineering school in the United States. Understandably, it had a strong curriculum in geology from its beginning and many of Eaton's students became prominent in the early state surveys, especially Ebenezer Emmons (1799–1863), James Hall, and Douglass Houghton (1809–1845).

CHAPTER ELEVEN

The State Surveys

NORTHERN CONSERVATIVES AND SOUTHERN LIBERALS

The philosophical and religious attitudes of American geologists in the first third of the nineteenth century show a definite regional divergence: The North was conservative and the South was liberal. The Yankees became Wernerians almost without exception and they took that position under the influence of their religion. Cleaveland, Dewey, Emmons, Silliman, and his students Hitchcock and Dana all followed a more or less literal interpretation of Genesis, saw Divine Harmony in nature, and tried earnestly to reconcile their geological observations with Mosaic history. They did not consciously warp geology for the sake of Genesis, but they struggled mightily to fit it into the biblical mold. In England, William Buckland had been the last serious geologist to try such a thing (Buckland, 1836), but in America these attempts continued for another half century. The struggle widened after 1859 when Darwin's *On the Origin of Species* appeared, triggering the religious war on evolution that still goes on. Geology, and particularly geologic time, became part of that debate. Among American geologists the most prolific contributor to that movement and an earnest searcher for parallels between Genesis and geology was Alexander Winchell (1824–1891). He was a Connecticut Yankee and long-time Professor of Geology at the University of Michigan (1853 to 1873 and 1879 to 1891), who also taught briefly at Syracuse and joined newly established Vanderbilt University in Nashville in 1875. He wrote a series of flowery popular books (Winchell 1870, 1874, 1877, 1880, 1881, 1883, 1886) and thought he was being successful in working out the conformation of both geology and evolution with Genesis. His Methodist trustees at Vanderbilt took a different view: They decided that he was warping Genesis to fit geology. In 1878 they abolished his chair and he returned to the University of Michigan.

Of the northern geologists, only Amos Eaton and James Hall avoided taking religious positions on geological questions. Eaton encouraged religion in others, but never joined a church himself. Hall preferred to keep his religious feelings strictly private, although it is known that he leaned toward Roman Catholicism for some time, and the Old Testament clearly moved him much less than it moved his devout Protestant colleagues. He could not abide the thought of taking time for anything other than collecting more fossils, describing them in the best way possible, and coercing the legislature to publish the results. He refused categorically to participate on any level of the great debate over Darwin's new theory, and "went on heaping up new facts to the end" (Clarke, 1923, p. 508).

The southern group was centered in Philadelphia and was led by Conrad, Vanuxem, and the Rogers brothers. They would not let the Noachian Deluge affect their geological thinking and felt free to ignore Wernerian concepts of the sequence of layered rocks. The early success of their empirical fieldwork was not lost on their northern colleagues, and was the main reason for the ultimate abandonment of Wernerian criteria in spite of their religious appeal. By the time the New York survey volumes began to appear in 1842, a new empirical American stratigraphic nomenclature was firmly in place, and the concepts of "Transition" and "Floetz" were abandoned.

The geographic distances separating American geologists were often large, but they managed to know each other and keep in touch. In 1819, in New Haven, they founded the American Geological Society, which held meetings until 1828. In 1840, in Philadelphia, they formally organized the Association of American Geologists which seven years later changed its name to the American Association for the Advancement of Science and has held regular annual meetings since then. In their purposes these organizations were quite different from their earlier British counterparts. American geologists organized to create a forum for the exchange of scientific information and a basis for professional recognition without regard for social status and with little money to spend on formal activities. Any active geologist was welcome as a member, and considering the enormous differences in backgrounds, resources, and temperaments among the group, it may seem surprising that they got along so well together professionally and that their scientific intercourse was so productive. From casual exchange of views at the annual meetings, whether cool or heated, to extensive cooperation and collaboration in the field, they forged the dramatic growth in the quality of American geology from about 1820 to about 1845.

This was the time of rapid growth for coal mining in Pennsylvania. In the bituminous coal regions new mines were opening wherever a means of transportation could be arranged. The Lehigh Coal and Navigation Company was incorporated in 1822 and rapidly developed the anthracite strip mines west of Mauch-Chunk and the transport to ship the coal to markets. Its predecessor companies had tried to sell the "hard coal" since about 1792, but it took more than a generation for the public to learn that a grate was required to burn it. Once that simple technology caught on, mining towns mushroomed in Pottsville, Tamaqua, Wilkes-Barre, and many other places.

A canal-building fever erupted after the Erie Canal opened in 1825, and the newly available means of bulk transportation created enormous business opportunities. A network of coal-carrying canals and gravity-operated railways was built to cope with the mountainous terrain of the coal fields. Coal-fired steamers began plying the inland rivers and Lake Erie, population was moving westward, and the sites of old trading posts were becoming cities.

A Brief Account of the Discovery of the Anthracite Coal on the Lehigh, by Thomas C. James, MD, (Memoirs of the Historical Society of Pennsylvania, v. 1, pp. 315–320, 1826.)

It was some time in the autumn of 1804 that the writer and a friend (*Anthony Morris, Esq.) started on an excursion to visit some small tracts of land that were joint property on the river Lehigh in Northampton county. We went by the way of Allentown, and, after having crossed the Blue Mountain, found ourselves in the evening unexpectedly bewildered in a secluded part of the Mahoning Valley, at a distance, as we feared, from any habitation; as the road became more narrow, and showed fewer marks of having been used, winding among scrubby timber and underwood. Being pretty well convinced that we had missed our way, but, as is usual with those who are wrong, unwilling to retrace our steps, we nevertheless checked our horses about sun-setting, to consider what might be the most eligible course. At this precise period, we happily saw emerging from the wood, no airy sprite, but, what was much more to our purpose, a good substantial German-looking woman, leading a cow laden with a bag of meal, by a rope halter. Considering this as a probable indication of our being in the neighbourhood of a mill, we ventured to address our inquiries to the dame, who in a language curiously compounded of what might be called high and low Dutch, with a spice of English, made us ultimately comprehend that we were not much about a mile distant from Philip Ginter's mill, and as there was but one road before us, we could not readily miss our way. We accordingly proceeded, and soon reached the desired spot, where we met with a hospitable reception, but received the uncomfortable intelligence that we were considerably out of our intended course, and should be obliged to traverse a mountainous district, seldom trodden by the traveller's foot, to reach our destined port on the Lehigh, then known by the name of the *Landing*, but since dignified with the more *classical* appellation of Lausanne. We were kindly furnished by our host with lodgings in the mill, which was kept going all night; and as the structure was not of the most firm and compact character, we might almost literally be said to have been rocked to sleep. However, after having been refreshed with a night's rest, such as it was, and taking breakfast with our hospitable landlord, we started on the journey of the day, preceded by *Philip*, with his axe on his shoulder, an implement necessary to remove the obstructing saplings that might impede the passage of our horses, if not of ourselves; and these we were under the necessity of dismounting and leading through the bushes and briars of the grown-up pathway, if pathway had ever really existed.

In the course of our pilgrimage we reached the summit of the Mauch-Chunk Mountain, the present site of the mine or rather quarry of Anthracite Coal; at that time there were only to be seen three or four small pits, which had much the appearance of the commencement of rude wells, into one of which our guide descended with great ease, and threw up some pieces of coal for our examination; after which, whilst we lingered on the spot, contemplating the wildness of the scene, honest Philip amused us with the following narrative of the original discovery of this most valuable of minerals, now promising, from its general diffusion, so much of wealth and comfort to a great portion of Pennsylvania.

He said, when he first took up his residence in that district of country, he built for himself a rough cabin in the forest, and supported his family by the proceeds of his rifle, being literally a hunter of the backwoods. The game he shot, including bear and deer, he carried to the nearest store, and exchanged for the other necessaries of life. But, at the particular time to which he then alluded, he was without a supply of food for his family, and after being out all day with his gun in quest of it, he was returning towards evening over the *Mauch-Chunk* mountain, entirely unsuccessful and dispirited, having shot nothing; a drizzling rain beginning to fall, and the dusky night approaching, he bent his course homeward, considering himself as one of the most *forsaken* of human beings. As he trod slowly over the ground, his foot stumbled against something which, by the stroke, was driven before him; observing it to be *black*, to distinguish which there was just light enough remaining, he took it up, and as he had often listened to the traditions of the country of the existence of coal in the vicinity, it occurred to him that this perhaps might be a portion of that *"stone-coal"* of which he had heard. He accordingly carefully took it with him to his cabin, and the next day carried it to Colonel Jacob Weiss, residing at what was then known by the name of Fort Allen. The Colonel, who was alive to the subject, brought the specimen immediately with him to Philadelphia, and submitted it to the inspection of John Nicholson and Michael Hillegas, Esqs. and Charles Cist, an intelligent printer, who ascertained its nature and qualities, and authorized the Colonel to satisfy Ginter for his discovery, upon his pointing out the precise spot where he found the coal. This was done by acceding to Ginter's proposal of getting through the forms of the patent-office the title for a small tract of land which he supposed had never been taken up, comprising a mill-seat, on which he afterwards built the mill which afforded us the lodging of the preceding night, and which he afterwards was unhappily deprived of by the claim of a prior survey.

Hillegas, Cist, Weiss, and some others, immediately after, (about the beginning of the year 1792,) formed themselves into what was called the "Lehigh Coal Mine Company," but without a charter of incorporation, and took up about 8 or 10,000 acres of, till then, unlocated land, including the Mauch-Chunk mountain, but probably never worked the mine.

It remained in this neglected state, being only used by the blacksmiths and people in the immediate vicinity, until somewhere about the year 1806, when William Turnbull, Esq. had an ark constructed at Lausanne, which brought down two or three hundred bushels. This was sold to the manager of the Waterworks for the use of the Centre-Square steam-engine. It was there tried as an experiment, but ultimately rejected as unmanageable, and its character for the time being *blasted*, the further attempts at introducing it to public notice, in this way, seemed suspended.

During the last war, J. Cist, (the son of the printer,) Charles Miner, and J. A. Chapman, tempted by the high price of bituminous coal, made an attempt to work the mine, and probably would have succeeded, had not the peace reduced the price of the article too low for competition.

The operations and success of the present Lehigh Coal and Navigation Company must be well known to the society; the writer will therefore close this communication by stating, that he commenced burning the Anthracite Coal in the winter of 1804, and has continued its use ever since, believing, from his own experience of its utility, that it would ultimately become the general fuel of this, as well as some other cities.

T.C.J.

Philada. April 13th, 1826

The following shows the quantity of coal sent from Mauch-Chunk to Philadelphia by water in the years specified, viz:—

In 1820 . . . 16,000 bushels.
1821 . . . 32,000 do.
1822 . . . 80,000 do.
1823 . . . 230,000 do.
1824 . . . 500,000 do.
1825 . . . 516,236 do.

In half the season, up to August 10th, 1826, there descended to Philadelphia 20,260 tons, equal to 567,280 bushels, which is a greater amount by 51,044 bushels than descended in the *whole* of the year 1825.

The great geological surveys of Pennsylvania and New York evolved in that economic climate. In Philadelphia a group of scientific-minded and business-conscious citizens, headed by Peter A. Browne (1782–1860), lawyer, promoter, and amateur geologist, chartered the Geological Society of Pennsylvania in 1833, with the stated purpose of promoting a geological survey. They lobbied vigorously until the authorizing act was passed in March 1836, and then disbanded. In New York the organizational lines had been less distinctly drawn, but the lobbying was just as intense and the act passed two weeks after Pennsylvania's.

ON-THE-JOB TRAINING

The formal education of most of the new survey geologists was sketchy: perhaps two years of college, with a course or two in chemistry and mineralogy, and only a few lectures in geology. They were a green lot when they started out, but as they worked over the often large territories assigned to them, they learned their profession the hard way—and fast. They soon picked out the regions geologically most interesting in the scientific and economic priorities of the time, and within a few years they had worked out the stratigraphic sequence well enough to be able to tell their state legislatures exactly where coal could and could not be found.

Lardner Vanuxem was perhaps the only one among them with rigorous training. He was a Philadelphian of French ancestry and had studied at the *École des Mines* in Paris, 1816 to 1819. Then he taught at South Carolina College until he quit in 1826, apparently in disgust with the state politicians, and followed the career of a free-lance geologist. He was the first to use fossils for stratigraphic correlation in New York and New Jersey, and showed that the Wernerian "Alluvial" of New Jersey was in fact Cretaceous, and correlative with the European "Chalk," in spite of the radical difference in landscape and lithology. He was also the first to present a coherent analysis of the flat-lying "Secondary" strata of western New York and to show that they are much older than Maclure and Eaton had supposed.

It was Vanuxem who proposed a new stratigraphic nomenclature based on local geographic names, with the idea of type locality clearly implied, and without any preconceptions of age. The Rogers brothers had begun by numbering their strata and then gradually developed a "universal" terminology based on Latin-derived expressions for the time of day, from earliest to latest: Primal, Auroral, Matinal, Levant, Surgent, Scalent, Pre-meridian, Meridian, Post-meridian, Cadent, Vergent, Ponent, Vespertine, Umbral, and Seral. It was a clever scheme, definitely not Wernerian, but contrived, rigid, and impractical when it came to naming formations that were being mapped in the field, but whose ages had not yet been determined. The Rogers brothers were fond of the idea and kept tinkering with the system, but did not properly formalize it until 1858 (Rogers, 1858), and by that time it was much too late to give it any chance of universal adoption. Their colleagues outside Pennsylvania had never been impressed. Vanuxem's system was used in New York, vociferously defended by Hall, and almost universally adopted. It still forms the basis of American stratigraphic practice.

THE "TACONIC QUESTION"

The sedimentary rocks of New York lie flat or nearly flat except in the easternmost part of the state, mainly east of the Hudson River, where they are severely folded, thrust, and metamorphosed. The northern part of that altered strip happened to fall into the Second District of the New York survey, assigned to Ebenezer Emmons, and he decided to make a detailed study of these puzzling formations all the way south to the Taconic Mountains, in the area where New York borders on Connecticut and Massachusetts.

Influenced by the Wernerian criteria he had learned as a student with

Dewey and later Eaton, Emmons concluded that these steeply-dipping, hard, and scantily fossiliferous strata must be older than the gently-dipping Potsdam Sandstone, then generally accepted as the oldest Cambrian formation in New York (now known to be Late Cambrian). He called these strata the "Taconic System" (Emmons, 1842, 1844, 1846), in spite of opposition from his colleagues, and especially Hall who had been his assistant in his first season (in 1836). Hall and Emmons had had their differences since their days together at the Rensselaer Institute and now it intensified (Clarke, 1923, p. 57); thus began the controversy over the "Taconic question," which polarized American geologists for two generations and continued to boil long after both Emmons and Hall were dead. (For a good critical review see Merrill, 1924, pp. 594–614).

Very little was known about Cambrian fossils when Emmons first proposed his Taconic System and when Hall first opposed it. Murchison's fight with Sedgwick was just beginning and Barrande's first paper on Bohemian trilobites appeared only four years later (Barrande, 1846). Practically nothing was known about the structures of metamorphic rocks or the field methods useful for studying them. Understandably, Emmons had mistaken slaty cleavage for bedding, and could not have easily recognized that parts of the Taconic section were overturned. Without detailed topographic maps he could not have successfully worked out either the structure or the stratigraphy, nor could his opponents prove conclusively that his field interpretations were wrong. It was just a matter of intuitions, and considering the rigid positions both sides were taking, plus the lingering animosity between the two principals, that made for a long, unpleasant fight but not one to bring much light to American geology. Emmons' Taconic System is now known to have included a multiplicity of Cambro-Silurian metasediments scrambled by tectonic movements in the complicated Appalachian mountain-building process.

LYELL'S FIRST VISIT

By 1841 the surveys of Pennsylvania and New York were nearing completion. They had been the most significant of all the state surveys in amount of geological exploration of the regions, extent of state support, and quality of work done. The preliminary reports which had appeared were necessarily addressed to a lay public, and some were heavily padded with trivial descriptions and statehouse double-talk, but the most important scientific results had been discussed at the geologists' annual meetings, and many were published in the *American Journal of Science*. The writing of the final reports had only begun, much important scientific information was still in preliminary form, and many loose ends remained. In Pennsylvania and New York the "last" field season was just beginning when word arrived through Silliman that Sir Charles Lyell was coming for an extended visit.

It was an awesome prospect. Lyell's *Principles* had seen six editions in London and a pirated one in Philadelphia (Lyell, 1837). Anyone could see that he had been everywhere, seen everything, and met everybody—on his side of the Atlantic. His talent for synthesizing other men's ideas was obvious from his book and no one in America could approach his stature as geologist, or so it seemed. The thought of having this "gimlet-eyed Englishman" peering over their shoulders as they struggled to finish their reports caused widespread discomfort among the young surveyors. It was an awkward moment in American geology, but the survey geologists could not help feeling honored by his obvious interest.

Sir Charles was fascinated by steamships and in July 1841, he took the fastest he could find, the packet *Acadia*. It whisked him and Lady Lyell from Liverpool to Boston in just 12½ days, including a day's stop in Halifax. He had been invited to give a dozen lectures at the Lowell Institute in Boston, and that was the main excuse for the trip, but he was intent on seeing all he could see of American geology, and to do justice to that goal he decided to stay a whole year. At the same time he was well aware of the potential for book sales in America, and a good part of his baggage must have consisted of the matrices for printing plates of the sixth edition of the *Principles* and the second edition of the *Elements*, which had just appeared in London. It was no accident that the first official American editions of those books "printed from the original plates and wood cuts, under the direction of the author" were published in Boston in 1841 (*Elements*) and 1842 (*Principles*).

Lyell went about his visit with all the single-minded determination he was known for and ultimately succeeded in arranging for himself a field trip through eastern North America, the like of which none had taken before and few since. He and his wife were on the road almost all the time, and they covered the region from Quebec to Savannah in trains, carriages, coastwise steamers, riverboats, and every other conceivable conveyance, usually guided by the best local people.

Lyell was genial, smooth, and convincing, but he had not known what to expect in America. He was aware that his reputation would serve as his passport with geologists and in the main centers of learning, and it did, but he was pleasantly surprised to find, out in the country, that a polite Englishman traveling for scientific purposes was warmly welcomed wherever he went and could count on all the assistance he could need, even from people who had never heard of geology. Travel in the backwoods of the United States turned out to be easier and much more pleasant than he had expected (Lyell, 1845).

Any qualms the young survey geologists may have had about being open with him gave way under the impact of his personality. Lyell was not a natural charmer, but he had faced this problem before and knew how to handle it. He praised them all to their faces and behind their backs. In his lectures he made favorable comments and handed out credit right and left. If he had expected American geologists to be half-trained bumpkins, he never expressed the thought, and if their recent accomplishments surprised him, he never showed the surprise. He even found nice things to say about Eaton's stratigraphy of the Niagara district and Percival's map of Connecticut! The few skirmishes that arose were effectively smoothed over with more praise and credit.

Among the survey geologists, Lyell had no trouble picking James Hall as the best thinker. Hall was cranky but he had vision and purpose—two qualities Sir Charles recognized and respected. The two became good friends for the rest of their lives.

The place Lyell most wanted to see in the United States was Niagara Falls, and that is where he headed two weeks after landing in Boston, guided by Hall. Lyell was then 44 and he had more than his share of experience in presenting his own point of view. Hall was only 30 and a beginner by comparison, but in the field (and on Hall's home turf) Sir Charles met his match. They spent five days together around the Niagara Gorge, and it must have been awesome to behold those two dynamos impressing one another with original geological observations. Among them, just to show that they were not getting everything right, was Lyell's conclusion that all the drift that covers the area was of marine origin. Agassiz had been to Scotland with Buckland the previous summer, and they had recognized moraines and other glacial effects almost everywhere. Buckland thought he had convinced Lyell that glaciers were very important in shaping the landscape, but now Lyell saw some recent marine shells in one of the elevated beaches along Lake Ontario and was thinking again in terms of marine wave erosion (Lyell, 1845, p. 37), unaware that the shells testified only to a brief interglacial marine incursion into Lake Ontario.

Hall showed him not only Niagara Falls but also the whole stratigraphic succession along the way from Albany. The American nomen-

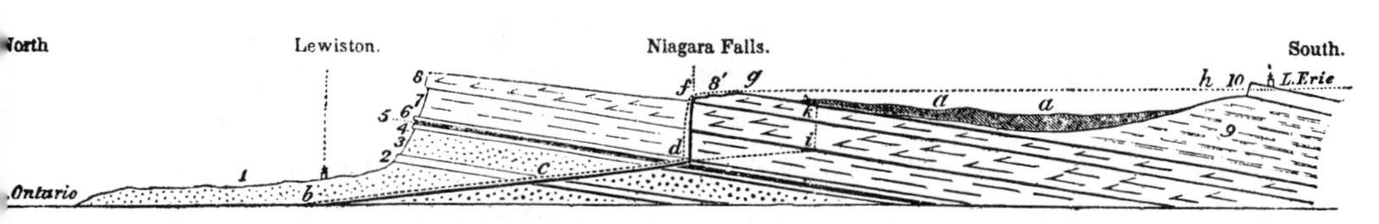

Section of the strata along the Niagara River, from Lake Ontario to Lake Erie.—(Chiefly from Hall's Report on the Geology of New York.)

FIGURE 11.1. Lyell freely borrowed from—and gave credit to—geologists who accompanied him to American localities. (Lyell, 1845)

clature was bewildering to Lyell, but Hall's mastery of the stratigraphy was obvious. As they traced the flat-lying formations across upper New York State, Hall pointed out the lateral changes they were observing in the lithology and in the fossils of continuous formations; it was clear that the sea in times long past had not been the same everywhere at the same time.

Amanz Gressly (1814–1865) had recently established the concept of facies in the Swiss Jura (Gressly, 1838), and Murchison and Sedgwick's Devonian (Rudwick, 1979) was, in fact, a new facies of the Old Red Sandstone, but it was Lyell's first chance to observe such transitions in the field on such a scale. "We must turn to the New World if we wish to see in perfection the oldest monuments of the earth's history, so far at least as relates to its earliest inhabitants," he wrote (Lyell, 1845, p. 15).

The timing of the visit was a fortunate coincidence and its effect was important on both sides of the Atlantic. The Geological Society of London was still the world's most respected geological forum; now its members were to be told by their busiest representative that there was important geology to be seen in America and American geologists were to be taken seriously.

On the American side, the man's ubiquitous presence called attention to his books. The *Principles* was far more substantial than Bakewell's widely used but trivial text, and the *Elements* presented the new biostratigraphy. The *Travels in North America* appeared in 1845 in London, was promptly reprinted in New York, and widely read. It presented the American geologic scene in very favorable light. American geologists had gone out of their way to accommodate their indefatigable visitor in his travels, and now he richly repaid them for their trouble. The self-respect they could derive from his blessing was oil for the lamp of American geology.

STRATIGRAPHY AND COAL

Stratigraphy was then on the cutting edge of the earth sciences, and the stratigraphy of New York provided the framework upon which the sequence of geologic events was to be built in the other states and territories. From Vanuxem's conceptual insight through Hall's dynamic action, the superposition of Paleozoic sedimentary rocks was being developed and documented. The understanding of lateral variation of fossils and lithology in continuous strata based mostly on observations of the facies relations of the sedimentary rocks of upper New York State was a major advance. The ability of New York geologists to cajole the legislature into publishing the results on a grand scale and on schedule played a large part in the overall success of their endeavor.

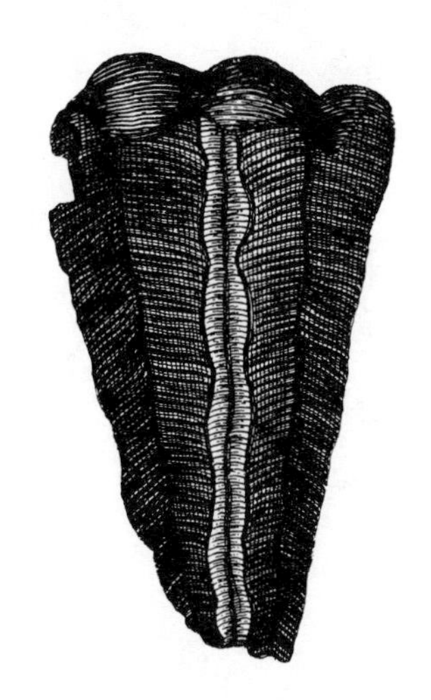

FIGURE 11.2. "Sept. 24*th*, 1819. —I send you four drawings of articles found by myself . . . No. 1 represents what is to me an incognitum; I do not find the like in Parkinson's 'Organic Remains,' nor in Sowerby's 'Mineral Conchology.' The drawing represents it exactly." (Letter from Caleb Atwater to Benjamin Silliman, 1820)

The Pennsylvania survey could have been just as important, but fate was against it (Lesley, 1876). State appropriations had run out in 1842 and no more support was in sight. That did not stop H. D. Rogers, who was not rich but knew how to manage and went right on doing what he thought was necessary. He was not an easy man to work for, but his choice of assistants was good and he knew how to maintain quality. By 1847 he transmitted a final report to the Commonwealth, but it took another 11 years before the *Geology of Pennsylvania* finally appeared in print (Rogers, 1858). It was a magnificent production, in keeping with Rogers' penchant for doing things right, was printed in Edinburgh because that was where he found the best printer, and was accompanied by the finest geologic map thus far made of an American state. Its impact, however, was weakened by the delays and by the conceptual cream-skimming that inevitably stemmed from them. The most impor-

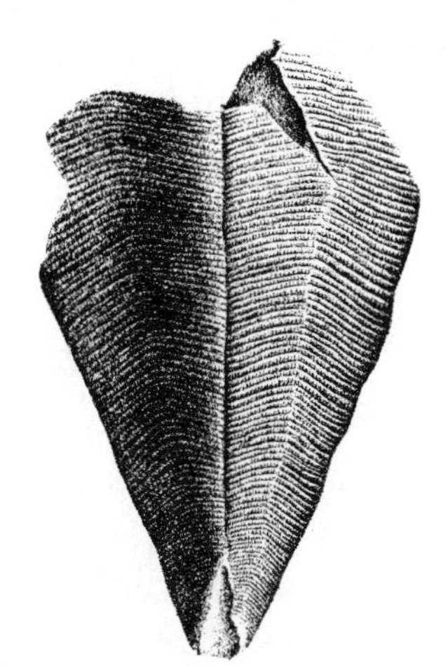

FIGURE 11.3. "*Conularia continens*, var. *rudis*. A specimen . . . showing the short, abrupt cone, and the strong surface-markings." (Hall, 1879, pl. 34A)

FIGURE 11.4. Plant fossils from the "Coal Measures" described by Lesquereux. (Lesquereux, 1858)

tant geologic ideas of the Rogers brothers, on Appalachian structure and on the origin and distribution of coal deposits, had been published long before, primarily in the *American Journal of Science.* The value of the Pennsylvania report was further diminished by H. D. Rogers' casual approach to invertebrate paleontology and by the insistent use of his contrived stratigraphic nomenclature long after it had been made completely obsolete by the wide acceptance of the New York stratigraphic system.

One of the high points of the Pennsylvania report was a monographic chapter on the coal plants of Pennsylvania by Leo Lesquereux, with 23 plates of accurate drawings. Lesquereux's name (1806–1889) runs like a common denominator throughout the survey reports of almost all states where plant fossils were found. He had come from a watch-making family in Neuchâtel in French-speaking Switzerland and had emigrated to America in search of political and religious freedom. He landed in Boston in 1848, spent some time with his compatriot Agassiz, then moved to Columbus, Ohio, where he settled for the rest of his life. In time he opened a watch shop—which still exists—and from there he conducted his free-lance mail-order research on mosses, peat, and fossil plants with imagination and a meticulous skill befitting his watch-making tradition. For four decades he was *the* paleobotanist of the United States.

In the 18 years between the first publication of the New York survey and the onset of the Civil War, the state surveys continued to work over their regions, squabble with their legislatures, and fitfully publish their reports, largely unhampered by the rising political turmoil. In the spirit of the time there was a kind of virtue in observing nature and describing it for the common good. Almost everything these young men found in the field, especially in the West, was new to science. The urgency to describe it all and illustrate it for posterity was one of the mainsprings of their science.

Based in what was left of his father's utopian colony in New Harmony, Indiana, David Dale Owen (1807–1860) conducted surveys that covered most of the territory of present-day Indiana, Illinois, Iowa, Wisconsin, and Minnesota, plus the states of Kentucky and Arkansas, supported partly by the states but mostly by the federal government. The emphasis was on the lead mining districts of the upper Mississippi valley, iron and coal wherever they could be found, the hot springs of Arkansas, and the exploration and surveying of land with agriculture in mind. It was geology in the service of governments with few distractions.

Owen was a good boss, attracted good assistants, and started them on productive careers. Edward Travers Cox (1821–1907), J. Peter Lesley (1819–1903), Fielding Bradford Meek (1819–1876), Benjamin Franklin Shumard (1820–1869), George Clinton Swallow (1817–1899), and Charles Whittlesey (1808–1886) all got their early field training under Owen. Swallow had studied at Bowdoin College with Parker Cleaveland, Lesley at Pennsylvania with Rogers, and Cox in New Harmony with Owen, but Meek, Shumard, and Whittlesey had little or no formal training in geology. They went on to direct geological surveys of their own, leaving no doubt of Owen's capability as a teacher of field geology, but Owen's own work is almost entirely routine and descriptive.

Owen gives lively accounts of paddling up cool rivers in the wilderness of Minnesota and of exploring the blazing gulches of the *Mauvaises Terres*, the Badlands of Dakota Territory. "The drooping spirits of the scorched geologist are not permitted to flag," he wrote (D. D. Owen, 1852, p. 197). He sent his invertebrate fossils to Hall, his extensive vertebrate collections to Leidy, and his plants to Lesquereux. He was not interested in theories and asked few fundamental questions.

Another interesting figure in the early state surveys was Douglass Houghton (1809–1845). He had studied and briefly taught under Eaton at Rennselaer and then went west to the frontier town of Detroit. Through his personal charm, practical outlook, and far-ranging business ability he soon became a leading citizen, took much interest in education, and gave numerous public lectures in geology, chemistry, and natural history. When Michigan formally joined the Union as a state in 1837, he became a professor at the new University of Michigan and head of the state survey.

COPPER IN MICHIGAN

Houghton had been to the copper country of the upper peninsula of Michigan in the summer of 1831 and his brief report (Schoolcraft, 1834, pp. 287–292) correctly concluded that the source of all the detached masses of copper found in the region since prehistoric times was "the trap rock of Keweena Point." He returned to the upper peninsula with his new survey to follow up the results of his early reconnaissance and "in his annual report, presented to the legislature of Michigan, February 1, 1841, the great features of the country were sketched with a masterly hand, and the first definite information with regard to the occurrence of deposites of native copper in the rocks was laid before the world" (Foster and Whitney, 1850, p. 13). Houghton then organized a systematic survey of the geology of the upper peninsula to be conducted hand-in-hand with the ongoing federal land survey along the grid provided by the land surveyors. Their project was barely under way when he was caught in a snowstorm in a small sailboat on Lake Superior near Eagle River on the night of October 13, 1845, and drowned. It was probably the greatest loss American geology had suffered in decades.

State Geological Surveys Before 1840

		Authors*
1823	North Carolina	Denison Olmsted
1824	South Carolina	Lardner Vanuxem (?)
1830–33+	Massachusetts	Edward Hitchcock
1831–45	Tennessee	Gerard Troost
1833–40	Maryland	J. T. Ducatel & J. H. Alexander
1833–39	New Jersey	H. D. Rogers
1835	Connecticut	J. G. Percival C. U. Shepard
1835	Virginia	W. B. Rogers

1836	Maine	C. T. Jackson
1836–42	Pennsylvania	H. D. Rogers
1836–	New York	L. C. Beck Timothy Conrad James Hall Ebenezer Emmons W. W. Mather Lardner Vanuxem
1837	Ohio	W. W. Mather
1837–1841	Delaware	J. C. Booth
1837	Indiana	D. D. Owen
1837–41+	Michigan	Douglass Houghton, J. W. Foster, & J. D. Whitney
1839	Rhode Island	C. T. Jackson
1839	New Hampshire	C. T. Jackson

*For a complete bibliography see Hazen and Hazen, 1980.

CANADA

The oldest federal survey in North America is the Geological Survey of Canada. It was established as a provincial survey with the appointment of William Edmond Logan (1798–1875) in 1842 and became federal with the British North America Act, which ratified the Canadian Confederation in 1867. Logan had been successful in business and had written an important monograph on a coal field in Wales (Logan, 1842), giving paleontological evidence that the coal was formed in place. In Canada he greatly expanded his scope, and literally covered the geology of the eastern part of the country (Logan, et al., 1863) west to what is now the western border of Ontario. That was a monumental achievement, considering that most of the region was then a trackless wilderness.

Logan outlined the three major geological provinces in Canada: the ancient crystalline complex, which he recognized as composed mainly of metamorphosed sediments and called Laurentian; the younger but still very old sequence of metasedimentary and intrusive rocks that lie unconformably over the Laurentian and which he called Huronian; and the extensive fossiliferous Paleozoic rocks of eastern Canada in what is now Quebec and the Maritime Provinces.

Murchison adopted Logan's term *Laurentian* for some crystalline rocks in Scotland (Logan et al., 1863, p. 22), and Logan returned the compliment by following Murchison's dictum and using *Silurian* to include all fossiliferous rocks below the Devonian (Logan et al., 1863,

p. 20). In his stratigraphic subdivisions, however, he followed the New York system of locally derived names. The New York survey had accepted Sedgwick's Cambrian without making an issue of it (Hall, 1843, p. 20).

The great thrust fault that runs along the middle of the St. Laurence estuary, turns south along the Hudson River, and then veers off southwestward across Pennsylvania is now known as Logan's line. Against considerable opposition, Logan maintained that the basins of the Great Lakes have been scooped out by flowing ice, and (with Newberry) asserted that the great till deposits that cover so much of North America had been laid down by continental glaciers (Logan et al., 1863, p. 889).

In his *Geology of Canada* Logan illustrated and briefly described what he (and his friend James Hall) supposed to be a fossil (Logan et al., 1863, pp. 48–49) from the upper Laurentian Grenville marble of the Grand Calumet channel of the Ottawa River near Bryson, Quebec. That was the seed of a long and often quite unreasonable transatlantic controversy over the biogenic versus the inorganic origin of *Eozoon canadense*, but neither Logan nor Hall took any part in the debate. The "fossil" is now known to be the inorganic product of metamorphism in limestones. (For a succinct chronicle of the controversy see Merrill, 1924, pp. 564–578.)

In his far-flung survey, Logan had the assistance of two talented associates, Elkanah Billings (1820–1876), paleontologist, and Thomas Sterry Hunt (1826–1892), chemist. Billings had been a lawyer in Toronto and an amateur fossil collector until he formally joined the Survey in 1856. He did most of the early Canadian paleontology, with an occasional assist from Hall, and is remembered for his fundamental work on crinoids and cystoids.

Hunt, a Connecticut Yankee, was a brilliant analytical chemist and chemical theorist with an excellent grasp of geological conditions and an unbridled imagination. He had learned mineral analysis from the younger Silliman, joined Logan in 1846, remained with the survey until 1872, and then moved to MIT as Professor of Geology where he also served as chemist for Lesley's Second Survey of Pennsylvania. Belatedly, Hunt took Sedgwick's side in the Cambrian-Silurian controversy, and his long article on the subject (Hunt, 1872) still remains one of the best chronicles of that fight and of its implications for this side of the Atlantic.

Hunt was a flamboyant personality, given to pronouncements of spectacular hypotheses, and obsessed with defending them. He could be called the last neptunist because he insisted that all granitic and gneissic rocks formed hydrothermally from aqueous solutions emanating from the cooling Earth. His militant defense of that position (combined with an untidy marital life) kept him in trouble through the latter part of his career.

While working with Logan, Hunt also taught chemistry (1862–1868) at McGill University, where John William Dawson (1820–1899) was principal. McGill had been a neglected small college when Dawson took it over in 1855, but he soon built it into a major university. For that and other educational accomplishments he was widely honored (and knighted in 1883), but his scientific fame rests on the wide and imaginative research he did in paleobotany. Others had found fossil plants in Devonian rocks before him, but Dawson was the first to make a systematic study of them (Dawson, 1859) and to work out their evolution on dry land. He also brought order to the coal plants of Nova Scotia and their stratigraphy, studied the animals he found in the still-erect fossil hollow stumps there, and went on to describe Cretaceous and Tertiary plants of the Canadian west. His *Geological History of Plants* (Dawson, 1888) was a standard text for decades.

At the same time, it was Dawson who fought for the organic origin of *Eozoon* beyond all reason, insisting that it was a giant rhizopod like a foraminifer, who absolutely refused to accept the concept of continental glaciation, and who adamantly opposed Darwin's evolution. Even while he was assailing Darwin, some Darwinists were using his assertions about *Eozoon* to bolster Darwin's theory. It was in spite of Dawson that they lost out when the argument for the organic origin of *Eozoon* finally disintegrated.

CHAPTER TWELVE

California!

THE U.S. EXPLORING EXPEDITION

Talk about mounting an American exploring expedition into the Pacific began in the 1820s. Whalers and sealers from New Bedford and Nantucket were sailing there in increasing numbers and the business was highly profitable. There was also trade to be found on the west coasts of North and South America as well as in the Pacific islands, and American sea captains were busy developing it. The British were consolidating their long-standing position in the western Pacific and the U.S. Navy, perenially dissatisfied by the amount of its appropriations from Congress, was looking for suitable peacetime action to justify requests for more ships.

The public mood was for expansion and it found an able spokesman in one Jeremiah N. Reynolds (1799–1858), a young lawyer from Ohio who was fascinated by exploration and had come forward in support of the "theory" that the Earth was hollow and open at the poles. That idea had been broached long before in Paris (Gautier, 1721), but how it found its way to Ohio is not clear; it was circulated too in 1818 by John Cleves Symmes, Jr. (1780–1829), "Late Captain of Infantry" (C. R. Hall, 1934, p. 139) as he was lobbying for an expedition to explore the holes at the poles. It was just as zany then as it is now, but it caught on as a subject for popular debate for a decade or more, and still appears in the literature of spiritualists and seers.

The recipe of economic growth, political expansion, and military ambition—with just a dash of popular pseudoscience for spice—took time to ferment, but it worked. The political bickering in Congress and in the navy went on and on, not only about the money and how large the expedition should be, but also what it should do and who should participate. Surveying shoals and coastlines (and showing the flag) were obvious jobs for the navy, but this expedition was to be something more than that, only there was no agreement on what *that* should be. Few of the promoters of the expedition harbored any thought of signing on themselves. Like the surveys of Pennsylvania and New York, the Exploring Expedition was being pushed "for the common good."

In the end, a technically minded and reasonably capable Navy lieutenant named Charles Wilkes (1790–1877) was placed in command and six ships were made available: two square-rigged sloops of war (the 780-ton *Vincennes* and the fast 650-ton *Peacock*), the new but slow 468-ton store ship *Relief*, the 230-ton brig *Porpoise*, and two New York pilot schooners (the 110-ton *Sea Gull* and the 96-ton *Flying Fish*). The schooners were selected for their supposedly very strong hulls; concern about the safety of the schooners on a long voyage was justified, as it turned out.

After many nominations, counter-nominations, and withdrawals, seven civilian scientists and would-be scientists finally signed on: William D. Brackenridge, 28, a nurseryman officially listed as horticulturist; Joseph P. Couthouy, 30, a former sea captain and enthusiastic shell collector; James Dwight Dana, 25, Yale's pride and a capable, if inexperienced geologist; Horatio Hale, 21, a recent Harvard graduate in philology interested in American Indian languages; Titian Peale, 38, artist and naturalist in Philadelphia and alumnus of several overland exploring journeys; Charles Pickering, MD, 33, Harvard-trained botanist and ethnologist; and William Rich, an amateur botanist who was a good friend of the Secretary of the Navy. Many others had wanted to go but were left behind, among them the one civilian who had done more than any other to keep the project alive, Jeremiah Reynolds. (For a lively account of the near-scandalous staff selection see Stanton, 1975).

The flotilla sailed from Norfolk August 18, 1838, rounded Cape Horn, crisscrossed the Pacific from what is now Wilkes Land in Antarctica to the northwestern coast of North America, and then returned by way of Manila, Singapore, and Cape Town to New York in June 1842. It was an arduous voyage. The *Relief* could not keep up and was sent home from Callao in July 1839 with a crew that included "all invalids and idlers," as Wilkes put it. The *Peacock* ran aground on the Columbia River bar and broke up in the surf, fortunately without loss of life. The pilot schooners fared badly: The *Sea Gull* disappeared with all hands in May 1839, off the coast of Chile, and the *Flying Fish* had to be sold in Singapore, after many repairs, to avoid the risk of sailing her home. Of the original crew of 346 men, only 181 returned with the expedition. Forty-six had deserted and about an equal number had lost their lives from various mishaps and diseases. At the end of their enlistment term, 48 men demanded discharge and obtained it in Honolulu. Not surprisingly, most of them were from the *Vincennes*, Wilkes' flagship. Three reluctant marines of the *Vincennes* were flogged in Wilkes' presence until they agreed to re-enlist. To replace the losses, 178 men were recruited in foreign ports, but almost half of them deserted before the end of the cruise (Stanton, 1975, pp. 279–280).

The personality of Charles Wilkes evokes a storm of contradictions and his own autobiography (Wilkes, 1978) is an excellent source. He was an undistinguished sailor, but he commanded a squadron under sail around the world (Tyler, 1968). He looked upon enlisted men as lower forms of life, but his survival literally depended on his enlisted crews. He knew perfectly well that his reputation would rest on the success of his Expedition and he spent most of his active life supervising and publishing the work of the largest corps of civilian scientists assembled in the federal service up to that time (Stanton, 1975), but he had no use for civilians and never hesitated to admit it.

As a commander Wilkes was rude, arbitrary, and overbearing—quick to find fault and slow to give credit. He was insecure, vain, and prone to regard any new idea as the seed of mutiny. He was no diplomat but succeeded in the Washington political scene. He often thought that "cabals" were forming to oppose him, but he always sprang back from every setback and survived being found guilty by two court martials. He certainly was no writer (Wilkes, 1844, 1978), but absolutely insisted on composing (or, more accurately, compiling) the five-volume *Narrative* of the Expedition himself (Wilkes, 1844). He padded it heavily with trivial textbook geography, copied verbatim from the journals he had confiscated from the officers and scientists under his command, and gave no credit. Even Wilkes' friends agreed that the *Narrative* was a literary disaster, but he obtained a copyright for it after the official edition had appeared, and personally arranged for the publication of several large unofficial editions that sold very well and had a great influence on American seafaring literature.

Wilkes had indomitable energy and a stolid military sense of duty, and he lived to complete the self-imposed task of organizing the collections and publishing the scientific results of his Expedition. Among the young officers of the U.S. Navy who had been willing to accept command of the Expedition in 1838, he almost certainly turns out to have been the best qualified.

The most productive civilian scientist of the Expedition was one of the youngest, James Dwight Dana. Officially he was the geologist, but when Couthouy was fired in a burst of Wilkesian mistrust, Dana also became the marine biologist. As fate would have it he was well prepared for both roles as a result of his stint as teacher to the midshipmen of the USS *Delaware* on a long Mediterranean cruise that took him, among other places, to Naples. He had learned how to live with the Navy and had gotten a good look at Vesuvius at the same time. Back at Yale he then produced his *System of Mineralogy* (Dana, 1837), which became the first essay of his magnum opus. It had five editions in his lifetime and is still in print. He was a lucid and organized writer and now, at the age of 25, he was already familiar with the mechanics of seeing a large and complicated book through the press. Those were powerful qualifications.

The squadron went south to chart the Antarctic coast in 1839. In Sidney, Dana read a newspaper report of the famous paper Charles Darwin had read in 1837 on the origin of atolls by the subsidence of extinct volcanoes (Darwin, 1842) and soon he was able to confirm Darwin's conclusions in Fiji. Then he made the important additional observation that the volcanic cones are eroded by running water and that the eroded shape, not the original shape of the volcano, determines the outline of the atoll.

Dana's geological results ultimately appeared in grand style in Volume X of the Expedition's official reports, with a folio atlas (Dana, 1849*a*). His work on the corals themselves became even more significant and produced the first systematic monograph on those "builders of the deep" (Dana, 1846), accompanied by another superb atlas (Dana, 1849*b*). That was Volume VII. Next Dana wrote two volumes (XIII and XIV) on Crustacea in which he described 500 new species (Dana, 1852–1853), with still another magnificent atlas (Dana, 1855).

At the very beginning of Wilkes' publication program Congress decided that the reports of this Expedition were to be something extra special. Only 100 copies were authorized to be printed. They were to be presented, one each, to the states of the Union, important foreign governments, and the commanders of the three large vessels of the Expedition. No copies were to be sold to the public and none were to go to the authors. It was not an economy move because the cost of additional copies would have been negligible once the type was set and the plates engraved, but all efforts to change that strange rule were thwarted. Administrations changed, but Congress insisted to the end that the official reports of the Exploring Expedition were to be rare books.

What the states and the foreign governments did with those bibliographic rarities was their business. Some deposited them in state universities and public libraries. The copy sent to the Emperor of China turned up in a Canton market stall in 1858, and when the Czar's set was lost at sea, special permission of Congress was required before he could be sent another (Stanton, 1975, p. 351). The scientists were furious, and a lot of good it did them. Dana had 100 extra copies of his volumes printed at his own expense and distributed many of them to libraries where he thought they would be made available, but the overall impact of the scientific work of the Expedition was obviously impaired by this curious contention of Congress.

A Summary of Federal Expeditions and Surveys Before 1840

1803–06	Meriwether Lewis and William Clark to the Pacific Northwest
1818–19	H. R. Schoolcraft to Missouri
1819–20	S. H. Long, Thomas Say, and Edwin James to the Rocky Mountains
1820	Schoolcraft to the sources of the Mississippi River
1821	Schoolcraft to the central Mississippi Valley
1823–25	Long, Say, and W. H. Keating to the upper Mississippi
1834	G. W. Featherstonhaugh to the Ozark region
1838–42	Charles Wilkes around the world (J. D. Dana, geologist)

GOLD

After the wreck of the *Peacock*, members of the crew, including most of the Expedition's "scientifics," spent several months ashore in the Pacific Northwest in the fall of 1841; they made an overland journey of more than a month from what is now Portland, across the Klamath Mountains, and down the Sacramento River to Yerba Buena (now San Francisco). Toward the end of that trip they came across the ranch of a thriv-

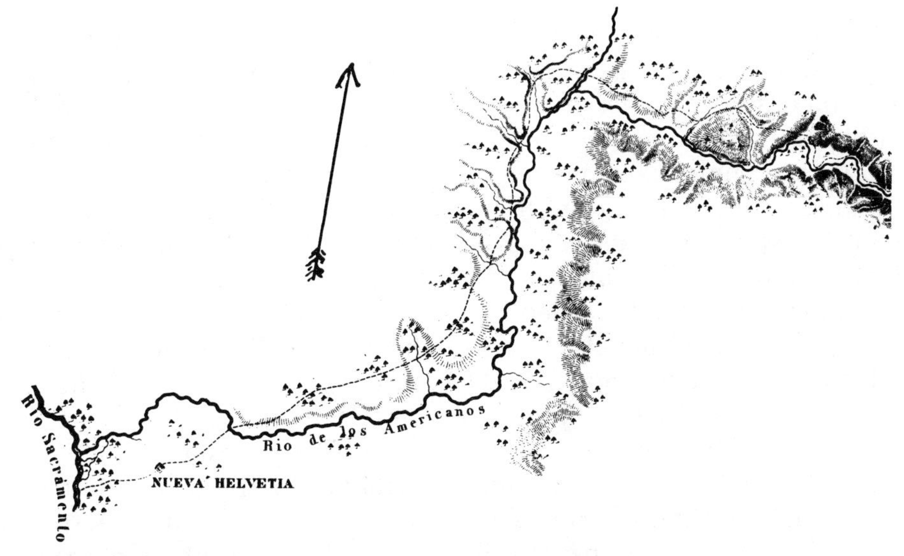

FIGURE 12.1. Frémont and, before him, members of the Wilkes expedition, never noticed the gold near Sutter's settlement. (The arrow shows magnetic north; Frémont, 1845)

ing pioneer of Swiss-German origin who had been there less than three years, but already was running several thousand cattle, horses, and sheep, and building a major colony of settlers under a land grant from the Mexican government. He was Johann Augustus Sutter (1803–1880), and the place he was building on the American River where Sacramento now stands was New Helvetia. As he entertained the party in his unfinished fort, little did anyone realize that the river below was rich with gold. Had Dana gone down to the stream with a frying pan to try the gravel, what a surprise he could have had!

Frémont's Camp Near Sutter's Settlement

March 4 [1844] We continued rapidly along on a broad plainly-beaten trail, the mere travelling and breathing the delightful air being a positive enjoyment. Our road led along a ridge inclining to the river, and the air and the open grounds were fragrant with flowering shrubs; and in the course of the morning we issued on an open spur, by which we descended directly to the stream. Here the river issues suddenly from the mountains, which hitherto had hemmed it closely in; these now become softer, and change sensibly their character; and at this point commences the most beautiful valley in which we had ever travelled. We hurried to the river, on which we noticed a small sand beach . . . We traveled on over the river grounds, which were undulating, and covered with grass to the river brink. We halted to noon a few miles beyond, always under the shade of the evergreen oaks, which formed open groves on the bottoms. (Frémont, 1845.)

Every river in New England had been panned for gold in the seventeenth and eighteenth centuries with uniformly negative results. A little gold, including a few large nuggets, had been found in Georgia and the Carolinas (already mapped by Guettard in 1752), but all attempts at mining it commercially were so discouraging that by Dana's time no thinking person would want to try it anymore. There was obviously no point in exploring American rivers for gold. The gold pan is not mentioned anywhere in Dana's or Wilkes' accounts of the Expedition, and only a last-minute note in *Geology* (Dana, 1849*a*) records the finding of gold by Sutter's partner at the sawmill he was building near what is now Coloma on the American River. Only after that discovery did the gold pan become standard equipment for western explorers.

The parallel between Darwin's voyage in the *Beagle* and Dana's in the *Peacock* has often been drawn and the similarities are many: Both men were in their twenties, both had an inquiring mind and a sharp eye, both were uniformitarians, and both studied marine life and coral islands. The great difference between them lay in their religious attitudes, and that turned out to overshadow all similarities. Darwin's outlook was liberal and strongly influenced by Lyell's *Principles*, where the break between Genesis and geology is clearly made. Dana had been raised on Silliman's edition of Bakewell's *Introduction*, where Creation and the Deluge are treated as physical facts and no contradiction is seen between geological observations and Mosaic history.

Shortly before he sailed with Wilkes, Dana had joined the archconservative First Church of Christ in New Haven, where literal interpretation of Scripture was a guiding principle. In the Pacific, wherever Darwin looked he saw a complex dynamic system, whereas Dana interpreted the same information as evidence of a static Divine Order. Hemmed in by his religious convictions, Dana was able to take an objective view of the development of atolls, but when he dipped his net into the sea and saw the unimagined variety of marine life, he could only describe it, classify it, and marvel at the vast complexity of Creation.

RAILROADS TO THE WEST

In March 1844, another exhausted young explorer appeared at hospitable New Helvetia—Brevet Captain John Charles Frémont (1813–1890) of the Corps of Topographical Engineers. Like Dana he had also taught mathematics to midshipmen just after leaving college, had become interested in surveying, and was Joseph Nicolas Nicollet's (1786–1843) assistant on two exploring journeys into the Dakota Territory. In 1841 he eloped with Jessie Benton, the very young daughter of the powerful senator from Missouri; Fremont wisely made peace with Benton, and Benton's support was essential in Frémont's subsequent explorations. He led his own expedition to the Rocky Mountains in 1842 and to Oregon in 1843–1844, and had come down the American River to New Helvetia from what is now Lake Tahoe, also without having seen any gold. Frémont and his wife were both good storytellers, and the day-by-day accounts of his travels make exciting reading (Frémont, 1845). He was keen to find new routes and passes, drew interesting maps, and knew his way not only through the Sierra, but in political Washington as well. Congress ordered 20,000 copies of his report to be printed in the same year that Wilkes was told 100 would be appropriate for him. (Frémont served briefly in the Senate, then ran for president in 1856, but was defeated by Buchanan.)

On his map Frémont had included a geologically annotated profile from the Green River to Fort Vancouver (now opposite Portland, Oregon). He had also collected a few fossils on the way, had sent them to Hall in Albany, and they were illustrated and briefly described in a 14-page appendix with five plates (Frémont, 1845). From the Muddy River (now Muddy Creek in the southwestern corner of Wyoming) Frémont brought back a piece of oolitic limestone and Hall wrote that in color and texture it "can scarcely be distinguished from specimens of the Bath Oolite" (now placed in the Middle Jurassic). That excited William Barton Rogers, and he wrote to his brother Henry (Mrs. W. B. Rogers, 1896, p. 265) on June 19, 1846: "I have lately been reading 'Frémont's Journal,' with his map before me, and felt an itching impatience to be able, united with you, to reap a portion of the great geological field west of the Rocky Mountains. From the very centre of Mexico, all the way to the Columbia, is a *terra incognita*, rich, it would seem, in geological phenomena. Must we not, some day, try our hammers on the Sierra Nevada, and trace the Oolite formation and others already reported of?"

The "itching impatience" may have been just another of William's phrases, but he was not alone in harboring such feelings. The West was beckoning, but inaccessible, and what little opportunity there was, even after the gold had been found, was reserved for the young men willing to rough it (and occasionally risk their necks) with the exploring expeditions. The danger from hostile Indians was very real and the distances between friendly settlements were large. Going to California was no small undertaking in the 1850s and early 1860s.

One could join a wagon train in Omaha or St. Louis and struggle across the plains and mountains for two or three months, or one could take a ship to Panama, ride through the jungle to Colon on the Pacific side, then take another ship to San Francisco. The trans-isthmian Panama Railroad opened in 1855, and that made the trip easier, but heavy goods still went around Cape Horn, a perilous route, particularly during the austral winter.

The northern boundary of the Oregon Territory was settled in 1846, California was ceded to the United States in 1848, gold was discovered later that year, and California became the thirty-first state in 1850. The pell-mell immigration of the forty-niners brought chaos to San Francisco, to Sutter's ranch, and to the burgeoning mining camps. The gold was there in large quantities, and some found it but most did not. Fortunately, the land was incredibly fertile and could support all the newly arrived population. As the euphoria subsided and reality dawned, California became settled in a hurry.

Philip Thomas Tyson (1799–1877), a Baltimore mineralogist, took a trip to California in '49, briefly studied the geology, and upon his return submitted a report to Frémont's old boss, Col. J. J. Abert, Chief of the Topographical Bureau. It was printed (Tyson, 1850) and gives a rough idea of California geology, with a conservative appraisal of the mineral resources. Tyson was trying to inject a voice of reason to temper the raging gold fever, but he greatly underestimated the amount of gold to be found in California. Next, a state-sponsored geological survey was attempted by John Boardman Trask (1824–1879) beginning in 1853. It issued several reports (Trask, 1853–1856), but the geology of California proved to be more than Trask could handle and more than the legislature was willing to pay for.

A survey finally began in earnest when Josiah Dwight Whitney (1819–1896) became state geologist in 1860. He had studied with Silliman and in Europe, had published a major book on American mineral resources (Whitney, 1854), and was an alumnus of the state surveys of New Hampshire, Michigan, Iowa, Illinois, and Wisconsin. He remained California's Geologist for 14 turbulent years, his aim being to sketch the geology of the state in fundamental terms and from a long-term point of view, without much regard for immediate returns.

Whitney tried to ignore pragmatic pressures and political winds, and that led to troubles with the legislature. In 1868 funds for his survey were not appropriated, but his father stepped in and funded the California survey from his own pocket. In time he was actually repaid by the state, but in 1869 the legislature passed the survey appropriations for only two more years and ordered all work to be finished by 1872. Whitney went right on working for two more years after that without any state support and then went back to Harvard as a professor for the rest of his days. One of the major contributions of his survey, *The Auriferous Gravels of the Sierra Nevada of California* (Whitney, 1880), was published by the Museum of Comparative Zoology of Harvard University without any help from the state of California.

The explosive growth of California generated a need for a better transport link with the industrial East and discussions of the possibilities for a transcontinental railway began soon after statehood was voted. The results of these explorations were published in a massive 13-volume report (War Department, 1855–1860) and included some important geological observations by Thomas Antisell (1817–1893), William Phillips Blake (1825–1910), Jules Marcou (1824–1898), and John Strong Newberry (1822–1892), among others. In later years, Blake became Professor of Geology at the College of California (now the University at Berkeley) and then director of the Arizona School of Mines at Tucson.

Newberry followed a broad career in geology and paleobotany, and in 1866 became professor at the School of Mines of Columbia College in New York. Antisell joined the Patent Office and dropped out of geology, but Jules Marcou, the prolific international storm petrel of American geology, had a vigorous altercation with Jefferson Davis, then Secretary of War, about the disposition of the fossil collections he had collected on the railroad survey. Marcou ignored his orders not to remove them from the United States, and many of his type specimens ended up in European museums. Marcou married into Yankee money and spent the rest of his life flitting between Washington, Paris, Zurich, and Cambridge (Mass.), usually surrounded by geological acrimony.

The political uncertainty preceding the Civil War delayed consideration of major western undertakings, but the war itself dispelled any remaining doubts about the usefulness of railroads. The construction of the Union Pacific Railroad began in 1865, heading west from Omaha, while the Central Pacific Railroad progressed eastward from Sacramento. They joined in Ogden, Utah, in 1869, and a wave of new traffic began moving between Chicago and California. The Atchison and Topeka Railroad, begun in 1859, became the Atchison, Topeka, and Santa Fe in 1863, but did not reach Santa Fe until 1880. The connections to southern California were completed only 10 years later. To the explorers, the railroads were an obvious convenience as well as a haven of safety from marauding Indians, and the railroad companies, with their large land holdings, looked more than favorably on geological surveys.

GEOLOGISTS IN THE FRATERNAL WAR

From the deepening division of the country, many had predicted secession and civil war, but few could have imagined the magnitude of the slaughter that finally ensued. Tremendous damage was done north and south, but the depredations on the lives of American geologists were surprisingly small. Oscar Montgomery Lieber (1830–1862), state geologist of North Carolina, was shot during the Confederate retreat from Williamsburg and died shortly afterward in Richmond. John Wesley Powell (1834–1902) lost his right arm at Shiloh in 1862 but remained in the army for the rest of the war. Ebenezer Emmons, who had become the state geologist of North Carolina, died there in 1863 (of natural causes) and his library and specimens were lost in the wartime turmoil.

James Duncan Hague (1836–1908) joined the navy in 1862 and served with the South Atlantic squadron a little more than a year. His brother Arnold (1840–1917) tried to enlist when he graduated from Yale but was rejected on physical grounds. Thomas Antisell and Ferdinand Vandeveer Hayden (1829–1887) dropped their geology for the duration and served as surgeons; Clarence Dutton (1841–1912), just out of Yale, joined the Ordnance Department, liked it, and came out of the war a Captain; Bache and Newberry were with the Sanitary Commission—all of them far from the shooting. James Hall only complained about the interruption of his fossil shipments from the South, but the political zeal and private anguish of the time are evident in the correspondence of the Rogers brothers, William in Boston and Henry in Glasgow, where he had held a chair since 1857.

The Rogers' brother-in-law, James Savage, was wounded at Cedar Mountain in 1862, captured, and died of his injuries in a Charlottesville hospital ten weeks later. William's letters had been heavy with patriotism, but when he tried to get to Charlottesville to see James, and returned unsuccessful but still hoping that James would recover, he wrote to Henry: "My visit to the field gave me a view of Bull Run, Manassas and other scenes of blood, and of the desolation which blasts this region in which the dreadful drama of war is enacted" (Mrs. W. B. Rogers, 1896, v. II., p. 131).

Most of the young geologists-to-be graduating from eastern universities must have had a good idea of that drama and they conspicuously avoided taking part in it. Samuel Franklin Emmons (1841–1911) received his A.B. from Harvard in 1861 and promptly sailed to Europe for prolonged studies. James Terry Gardner (he sometimes spelled it Gardiner) (1842–1912), his friend Clarence Rivers King (1842–1901), and their older friend Othniel Charles Marsh (1831–1899) all graduated from Yale together in that fateful year, 1862.

Marsh went to Europe, while King and Gardner roamed about until spring and then joined a wagon train to California, and in time both found gainful employment on Whitney's survey. Arnold Hague went to Germany and met Emmons while both were studying at Freiberg. Edward Drinker Cope (1840–1897) had not followed a formal curriculum, but he studied briefly with Joseph Leidy (1823–1891) at the University of Pennsylvania while classifying reptiles at the Philadelphia Academy of Natural Sciences and at the Smithsonian Institution in Washington. As a practicing Quaker he could not condone the war and in 1863 he went to Europe. Grove Karl Gilbert (1843–1918) graduated from the University of Rochester, also in 1862, and went to work for his geology teacher, Henry Augustus Ward (1834–1906), in his Natural Science Establishment. Raphael Pumpelly (1837–1923) went to Japan. F. B. Meek was a little too old for the war, and remained in his unsalaried position at the Smithsonian, describing fossils.

Matthew Fontaine Maury (1806–1873) was America's (and perhaps the world's) first physical oceanographer. He had been slated to be the Exploring Expedition's astronomer, but resigned in one of the many disputes, even before Wilkes was appointed commander. Later he succeeded Wilkes as the head of the Depot of Charts and Instruments of the Navy and was placed in charge of the new National Observatory (later the Naval Observatory) in 1844. In those positions he had access to a large collection of ships' logs, full of reports of winds and currents, and he realized their significance as a data base for navigation. In 1847 he published the first of his *Wind and Current Charts* (for the North Atlantic) and went on collecting more data, often by trading his charts for more information. He promoted and directed the first systematic program of sounding the North Atlantic, and presented the results in the first bathymetric chart of that ocean. His major work is the *Physical*

Geography of the Sea (Maury, 1855), which was instantly successful and was reprinted many times.

Maury was a great promoter and collector of empirical data, but his theories in meteorology and oceanography were basically unsound. He was severely criticized by A. D. Bache and Joseph Henry for insistently giving explanations for wind and current patterns that were contrary to physical evidence. His scientific reputation began to fade even before his famous book appeared, but his position in the Navy gave him ample opportunity for political exposure and he used it to the hilt. He was a Virginian, vocal about his southern sympathies, and in 1861 he joined the Confederate Navy. That, more than all the rebuke from his scientific adversaries in the Coast Survey, the Smithsonian, and the American Association for the Advancement of Science, terminated his research career in the Navy. He spent the rest of his life teaching physics at the Virginia Military Institute.

GEODESY

Among the earth sciences, perhaps the least spectacular, but probably the most difficult, has always been geodesy, the study of the figure of the Earth and its gravitational fields. From the beginning, it always involved very precise measurements with instruments on the very edge of contemporary technology, and large, difficult calculations to reduce the observed data. That is the main reason why geodesists have always been a small fraternity, but their work is fundamental to land surveying and mapping. Maps have always been important not only to commerce but also to the military, and that is why governments have been involved in geodesy from the beginning.

The government of the United States began thinking about a survey of American coasts almost as soon as it was first formed, but little was done until a newly arrived Swiss geodesist named Ferdinand Hassler (1770–1843) convinced President Jefferson and Treasury Secretary Gallatin to establish a Survey of the Coast in 1807 (Cajori, 1929). In 1811 he was sent to London to buy a repeating theodolite with a 2-foot circle, two transits, chronometers, standard scales, barometers, thermometers, and all the rest, was caught there by the wars, and stayed five years.

In 1816 coastal triangulation actually began in New York, but two years later it was suspended for lack of funds. In 1832 the Survey of the Coast was revived with Hassler again in charge, and it has continued ever since. In 1836 its name was changed to Coast Survey, and in 1878 Congress decided to call it the Coast and Geodetic Survey, belatedly recognizing the nature of the work. By the time Hassler died in 1843, the

Survey had carried a primary triangulation net from a baseline on Long Island to Rhode Island and to the Chesapeake Bay.

In Hassler's time a first-rate theodolite with a large achromatic telescope could measure horizontal angles with a precision better than 1 second of arc by careful repeating. Vertical angles in astronomical observations of latitude could be determined with portable transits to about 10 seconds, which corresponds to an uncertainty of about a quarter mile in the field, in a north-south direction. Measurement of longitude, which had been the subject of keen research and much debate in London in the eighteenth century, was still a problem in the 1830s. Accurate chronometers were available in England, but they were expensive and fragile. They would not have withstood being hauled in a wagon on American roads, and Hassler had a special carriage built to move his precious arsenal of instruments from station to station. A 2-second uncertainty on the clock was about the best that could be done, and that meant about half a mile in our latitudes.

Other methods of determining longitude were being examined, but none were successful. Alexander Dallas Bache set up the first American magnetic observatory in the garden of his house in Philadelphia in 1830, and aware that fluctuations in the direction of the magnetic needle could be related to longitude, he set up a much more elaborate one in a special nonmagnetic building at Girard College in 1840. The process was still far short of the accuracy required. Like Hassler, Bache was also very good with instruments, but when he came to Washington to succeed Hassler as director of the Coast Survey, he soon demonstrated another great talent: administration. He directed the Survey until his health failed 21 years later, and in that time it became the greatest scientific agency in Washington.

Field operations were greatly expanded. New baselines were established and the primary net extended all along the coasts of the United States, providing firm points of departure for secondary and local surveys. Telegraphy was introduced for time signals, thus dispensing with the clocks and their problems. The map of magnetic declination that Bache had constructed for Pennsylvania (Bache, 1863) was extended over much of the United States, and precise tide-recording stations were placed along the coast. Two of them (in San Francisco and in San Diego) were able to detect the arrival of very long "tidal" waves (now called tsunamis) from a Japanese earthquake in 1854 with sufficient certainty to permit calculating the wave-propagation velocity from Japan to California, and estimating from it the mean depth of the waters of the Pacific Ocean (Bache, 1856).

During this time, Bache became a powerful figure in the politics of American science. Together with his friend Joseph Henry (1797–1879), the great physicist and Secretary of the Smithsonian Institution, they became the voice of science in Washington and were largely responsible for the new relationship of the government with the scientific community that proved so successful after the Civil War. Bache served as the first president of the National Academy of Sciences, founded in 1863, but he did not live to see the most important product of his and Henry's political groundwork: the great western surveys.

CHAPTER THIRTEEN

The Great Western Surveys

The Civil War was over and the armies were going home, but the War Department was still the most powerful agency in Washington. Dozens of generals were searching for suitable employment, returning veterans were picking up where they had left off but with an eye out for new opportunities, and the nation was ready for expansion in the West. Geological exploration was seen as one means to that end, and the American geological establishment recognized the opportunity and seized it with great skill. The large amounts of money needed for the western surveys had to come from the Congress, but plans for the work and the continuous lobbying to keep it going came from the scientific community. Attitudes had changed since the time of the Wilkes expedition. Politics was far from dead, of course, but this time it played only very small parts in the scheme of appointments. Scientific voices prevailed, from Albany, Boston, New Haven, Philadelphia, and Washington, and highly capable young men were selected to make the surveys.

THE KING SURVEY

Clarence King (1842–1901) was the first in the field, in 1867. His sponsor was the U.S. Army, but his orders had a distinctly nonmilitary tone: " . . . to examine and describe the geological structure, geographical condition, and natural resources of a belt of country extending from the 120th meridian eastward to the 105th meridian, along the 40th parallel of latitude . . . " The orders were issued by General A. A. Humphreys, Chief of Engineers, but the text was written by Clarence King himself (Wilkins, 1958, p. 96).

King was then only 25 years old, but he radiated an intellectual promise that few could match and his senior colleagues stood solidly behind him. His fieldwork with Whitney in California and Nevada would have

FIGURE 13.1. Clarence King in California, about 1864. (From a photograph by Silas Selleck, San Francisco; courtesy of the U.S. Geological Survey)

FIGURE 13.2. Jim Gardner with his sextant, about 1864. (From a photograph by Silas Selleck, San Francisco; courtesy of the U.S. Geological Survey)

been qualification enough even without his demonstrated analytical insight and brilliant literary talent. His lifelong friend and critic, the historian Henry Adams, described him as the "best and brightest man of his generation" and it stands as a monument to the memory of General Humphreys, who remained King's superior throughout the six-year survey, that he was awed by this spectacular young man but not afraid of him.

For his staff King picked other young men of conspicuous talent: Samuel Emmons and the Hague brothers as geologists, James Gardner as topographer, and later Sereno Watson (1826–1892) as botanist and Ferdinand Zirkel (1838–1912) from Leipzig, the leading practitioner of the newly developed art of microscopic petrography. The King survey was a small operation, with a relatively low budget and rarely more than a dozen men in the field. It covered an enormous territory—almost 80,000 square miles. Its scientific results were presented in seven elegant quarto volumes with two atlases, and provided an insight into the geology and geography of their region on a level never before even approached.

In their last season in the field, a marvelous bonanza of publicity fell into their laps: the great diamond hoax. They sniffed it out and exposed it, and geology—their geology—became front-page news. No matter how good their work may have been, it never hurts authors to be known before they are published. The notoriety served them well and they used it with discretion. The victory of unselfish science over commercial fraud was a valuable lesson for the public and for Congress.

The Diamond Hoax and Clarence King

One day in the fall of 1871 two grizzled prospectors appeared at the Bank of California in San Francisco, handed over a little leather pouch to a teller and asked to have it stored in the vault. The pair, Philip Arnold and John Slack, "reluctantly" let the teller see its contents. It was filled with raw diamonds. They waited quietly for weeks while the news spread, as they knew it would. A financier of sorts, Asbury Harpending, and other investors negotiated a deal with Arnold to form a company to exploit the fabulous mine, its location still kept secret by Arnold. An exploring expedition guided by Arnold was led to a remote sandstone capped mesa in the northwest corner of Colorado—there they found diamonds, sapphires, and "rubies" (really mostly garnets) scattered over the rock, in cracks, on anthills.

In 1872, the U.S. Government Survey of the 40th Parallel, led by Clarence King, was finishing up its work. The young men of the survey had, of course, heard rumors of the great diamond find since it was in their territory. Curious as to the location of the mine and suspicious of the strange mineralogical association of the various stones, they figured out from the activities of the diamond company people and hints they had dropped, roughly where the mine was located. They mounted an expedition and found the mesa and "an association of minerals of impossible occurrence in nature." King returned to San Francisco to expose the hoax, a firm of London diamond brokers reported they had sold numerous stones to two Americans, and ultimately it was revealed that Harpending and Arnold had jointly concocted several previous mining-stock swindles.

(For a more detailed account of the diamond hoax, see: H. Faul, "Century-old diamond hoax revisited," *Geotimes*, v. 17, no. 10, pp. 23–25, 1972.)

THE HAYDEN SURVEY

Ferdinand V. Hayden's qualifications for geological exploration were, if anything, even more impressive than King's. He had lived and worked with James Hall in Albany while still in medical school from 1850 to 1853, and there he met Fielding B. Meek, a brilliant paleontologist and his future partner in exploration. Together they showed their ability as first-rate biostratigraphers of the badlands and bluffs of the Missouri valley.

Working for James Hall had not been easy. The old man was never satisfied and continually exceeded his budget. Hayden and Meek had to learn not only how to maintain the highest professional standards, but also how to subsist in the field with little or no money. It was not a stable arrangement, and before long Hayden was promoting his own support and Meek found a niche in the Smithsonian Institution. The war interrupted their fruitful collaboration, but as soon as it was over, Hayden was back seeking funds for more exploration. Nebraska became a state in 1867 and Hayden heard of an appropriation made to the General Land Office through the Department of the Interior for a geological survey of the new state.

Hayden was the obvious choice for the job because he knew the territory better than any other geologist at the time. The success of the survey was assured in advance because he and Meek, when together they had collected fossils there for Hall, had already done much of the stratigraphy that was specified to be done. Everybody was pleased and by 1869 Hayden was in charge of the newly constituted U.S. Geological Survey of the Territories, which was now directly under the Secretary of the Interior.

Hayden kept his sponsors happy by showing them what they wanted to see. His reports were usually optimistic, especially where agriculture and resources were concerned, even if occasionally slapdash and incomplete, but Congress loved them and Hayden's appropriations grew and grew. He had no trouble finding all the money he needed to attract the best talent and to publish their work in a style to which they had not been accustomed. From King's survey he brought the two best topographers, J. T. Gardner and A. D. Wilson, to produce his celebrated atlas of Colorado (Hayden, 1877), which is even better than the atlas of the 40th-parallel region which those same two worthies had prepared for King (1876).

Through his connection with the University of Pennsylvania, Hayden obtained the services of Joseph Leidy and his star pupil Edward Drinker Cope. Their paleontology required numerous fine lithographic plates and was published in three quarto volumes, one of them 5 inches thick (Cope, 1884)! When Leidy grew tired of vertebrate paleontology and disgusted with the running feud between Cope and Marsh, he returned to microbiology and Hayden published his large volume on freshwater rhizopods with 47 colored plates (Leidy, 1879). In spite of his friendship with John S. Newberry, his old friend from Oberlin College, Hayden instead engaged Leo Lesquereux for his paleobotany and published his work in three volumes with many more plates (Lesquereux, 1874, 1878, 1883). The cost of printing the work of all these virtuosi was enormous, but Hayden was able to cover it.

Hayden did not have the perception displayed by King, but then, who did? His annual reports may have left something to be desired, but the quality of his final ("permanent") publications was uniformly excellent. He was politician as much as he was geologist, and that led him into various scrapes with his colleagues who may have been purer in their science but poorer in their funding. Only Hayden had the gall and the drive to "discover" Yellowstone and to have it made into the first National Park in 1872, thus setting in motion the whole National Park movement.

Without Hayden, Leidy, Cope, and Lesquereux could not have produced their monumental contributions, yet it was the very intensity of Hayden's politicking that ultimately lost him the directorship of the U.S. Geological Survey when it was organized in 1879. At the head of the new Survey, the scientists wanted a scientist, not a politician, and Congress agreed.

THE U.S. GEOLOGICAL SURVEY

Clarence King was chosen, laid the plans, and set the new organization on its course (Rabbitt, 1979). He brought together a large group of talented geologists, including Dutton, Emmons, Arnold Hague, Gilbert, Hayden, Powell, Pumpelly, Walcott, young Bailey Willis (1857–1947), and A. D. Wilson, and set up a chemical department with Carl Barus (1856–1935) and William Francis Hillebrand (1853–1925). He had great ideas, but things were not going his way fast enough. The Organic Act of the Geological Survey had restricted its field of operations to the "national domain" and that was interpreted as the public domain. King wanted to study mineral resources in the whole country, not just the public lands of the West, but Congressional action would have been required for that, and Congress had other problems. James Garfield had been elected president by a very small majority in 1880, and the Senate was split down the middle, 37 Democrats and 37 Republicans. There was opposition to allowing the Survey to explore state and private lands, and King's old adversaries Hayden and Wheeler did their best to fuel it, but nothing was done anyway, and King resigned after less than two years in office. To succeed him, Garfield appointed a man whose credentials were then much less conspicuous than King's or Hayden's, but who turned out to have been the most able administrator of the three: John Wesley Powell.

FIGURE 13.3. Opposing viewpoints: John Wesley Powell (left) and Ferdinand Vandeveer Hayden (right). (Powell photograph courtesy of Pioneers Historical Soc., Flagstaff, Arizona; Hayden photograph courtesy of Denver Public Library Western Collection, W. H. Jackson, photographer)

THE POWELL SURVEY

Powell was the son of a Methodist preacher, an ardent abolitionist who traveled a lot and frequently moved his family. John Wesley went to school here and there, displayed unusual talent but did not find a channel for it, taught here and there, and was developing a reputation as a country naturalist in Illinois when the Civil War came. He enlisted and the army suited him. The combination of his abilities found response and in six months he was a lieutenant of artillery. His accomplishments in military engineering and fortification brought him to the attention of important commanders, including General Grant, and he remained in the army even after he had lost his arm. When the war ended, Major Powell returned to teaching in Illinois, this time at Illinois Wesleyan University, but his heart was not in it. He was obsessed with the idea of exploring the great western rivers and everything he did was aimed toward that goal.

In 1867 and 1868 he led parties of amateurs into Colorado, and then went to Washington in search of money for a major expedition down the Colorado River. The canyon of the Colorado was the last unexplored region of the United States, but even Powell's friendship with people in high places was not enough to generate an appropriation. The Smithsonian was very interested and everyone else offered encouragement, but money was not forthcoming. Some support finally came from several Illinois institutions, including two railroads, but a good part of the budget was covered by the major himself, from his own pocket.

His crew was a mixture of cronies, relatives, and adventurers without conspicuous qualifications. They carried surveying instruments, but Powell was the only scientist among them. They sailed in four boats from Green River, Wyoming (traveling to their starting point on the just-finished Union Pacific Railroad), on May 24, 1869, and almost everything they knew about rivers and boats they learned right there. They were heading for wild white water, but their equipment was woefully inadequate. They had only a few waterproof containers, and their food and clothing were inadequate. In retrospect we know that it was the beginning of a great geological survey, but at the time it could have been reasonably viewed as a reckless and pigheaded adventure.

Postwar America enjoyed wild stories from the West, some true but many not, and the major's expedition was a good subject. Less than a month after the boats left Green River, one John A. Risdon appeared in Omaha, reporting that Powell and his men had drowned in the rapids of the Green River and that he was the only survivor. Several newspapers picked up the story. It must have been a harrowing time for Mrs. Powell, but it was the luckiest possible break for the major. When his little flotilla appeared at the Ute Indian agency (in what is now Dinosaur National Monument), 37 days out of Green River Station, their "miraculous deliverance" was instant news, and the whole country now knew about the major and his expedition.

They continued down the river and braved enormous hazards. Two boats were lost, but no one was seriously injured. Near the end of the trip, three men became discouraged, left the party in the lower Granite Gorge, climbed out of the canyon, and were killed by Indians. Food ran out, clothes were in tatters, but when the major and his frazzled party finally arrived at Callville (a Mormon settlement now under the waters of Lake Mead), he was a national hero. They had spent more than three months in the canyon and accomplished little of geological value. Much of the survey data was lost, but Powell's name was now a household word, and his career as an explorer was established.

He made a second trip down the canyon in 1871 with another crew of friends and adventurers (this time without any loss of life), and finally got started on his systematic survey of the canyon country in the following year. It was only then that Powell's talent for scientific organization and his ability to stimulate first-rate thinkers were revealed. He brought in G. K. Gilbert from the Wheeler survey in 1874, and Clarence Dutton from the army in 1875. He boosted Clarence King for Director of the U.S. Geological Survey in 1879, and then took over himself when King resigned. Thus he helped form and then inherited a corps of unusual range and ability. Powell was the survey's director almost 14 years, and during his tenure it became the world's leading geological organization.

THE WHEELER SURVEY

The fourth great survey was organized by the army in 1871 and Lieutenant George M. Wheeler (1842–1905) was placed in charge, commanded by the same General Humphreys, Chief of Engineers. The army had finally realized that it was being squeezed out of its traditional role of map maker and planner of strategic communications by a group of fast-talking geological civilians. Wheeler had graduated sixth in the class of 1866 at West Point, but that was scant preparation for the job. The scientific establishment had never heard of him. Gilbert stayed with him for three seasons and Cope for one, but the main object of his survey was topographic mapping of the desolate terrain of Arizona and Nevada, according to his orders, and that was not exciting to his geologists.

Logistically it was a difficult job and there was trouble. Now it is hard to assess just how many of the problems reflected the commander's lack of judgment and how many were just bad luck. Wheeler's own report (Wheeler, 1889) is neither complete nor particularly candid, and somehow, he had developed a bad press. He understood the need for publicity, but chose the wrong people to handle it for him. Frederick W. Loring, a young Bostonian he had engaged to report the progress of the first field season, turned out to be an indifferent journalist. His dispatches often seemed to concentrate on the wrong things and the grandeur of the West was lost on him. Finally he was killed by Indians in the ambush of a California-bound stage, and that certainly was not the lieutenant's fault.

Gradually, the scientific establishment in Washington ganged up on the now Captain Wheeler, his appropriations dwindled, and when the U.S. Geological Survey was organized under scientific and patently civilian control, Wheeler and Hayden found themselves free to finish their projects, but the action was in the U.S. Geological Survey, and someone else was in command (Nelson, Rabbitt, and Fryxell, 1981).

THE REPORTS

The final scientific publications of the great surveys are impressive. (For bibliography and index see Schmeckebier, 1904.) King's group was first and set the tone with their *Mining Industry* (Hague, 1870), numbered "Volume III" of their projected series, then intended to run to five volumes. It is a milestone in the study of American mineral deposits, and more shall be said about it later. It has several chapters by James Hague, his brother Arnold, and S. F. Emmons, which detail the geological character of individual ore deposits in Nevada and Colorado and present engineering aspects of the mines exploiting them. These chapters give good insight into the explosive development of mining and ore dressing technology which was then under way and ultimately led to the United States becoming the world's leading supplier of mining and ore dressing machinery, with Denver and San Francisco the principal centers of the new industry.

The next volume of the King series was the first systematic monograph to describe the desert and dry-land flora native in the Great Basin (Watson, 1871; numbered "Volume V"). Sereno Watson (1826–1892) had appeared in King's camp near Pyramid Lake in western Nevada in July 1867, exhausted and half-starved, having just hiked across the Sierra from Sacramento. Watson was a strange fellow, and at first King did not quite know what to do with him, but his abilities soon became apparent, and when the survey's regular botanist, William Whitman Bailey (1843–1914), withdrew for health reasons the following year, Watson took over and soon became an important figure in western botany.

Early in the development of microscopic petrography, King realized the importance of the new art and brought Ferdinand Zirkel to New York in 1874, first to advise and then to take charge of the petrographic study of the survey's crystalline rocks. The result was "Volume VI" (Zirkel, 1876), the first comprehensive work on the petrography of

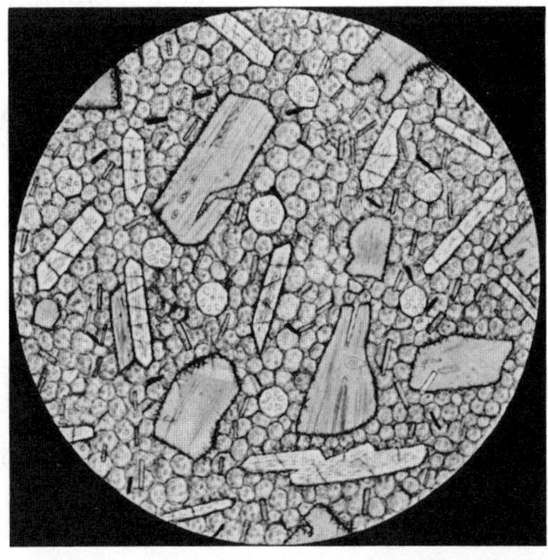

FIGURE 13.4. ". . . many rounded or eight-sided leucites, with grain-rings in their interior . . ." A thin section by Ferdinand Zirkel of a King Survery rock. (Zirkel, 1876, pl. 5, fig. 4)

American rocks. It presented many new rock types, including the first American occurrence of leucite (in the Leucite Hills, northwest of Point of Rocks, Wyoming), and opened a whole new field of research in this country.

A volume combining paleontology and ornithology came next (Meek et al., 1877), and then two thick summary volumes, one describing the regional geology district by district (Hague and Emmons, 1877) and the other taking a historical approach (King, 1878), sketching the geological events that had shaped the King survey's vast territory from earliest geologic time to the most recent. Here King did not present any great new geological concepts, but he covered an area that was geologically new and very different from the sections that had been studied in Europe, thus showing the world that geology was much more diverse than had been supposed. "Mr. King's graceful pen never showed itself to better advantage," wrote the *American Journal of Science* (in January 1879).

Finally, as "Volume VII" of what was to have been a five-volume series, King published O. C. Marsh's first major monograph (Marsh, 1880), describing a new kind of primitive bird with teeth. Marsh had not been a regular member of King's team, and some of the toothed birds had been collected in Kansas, far from King's assigned territory, but King saw the importance of the work because it proved the connection between birds and reptiles. He had some publication money left after the *Systematic Geology* (King, 1878) and its handsome atlas was printed. Marsh, after all, was a Yale man.

The final reports of the Hayden survey, also handsome and opulent, also had been planned in 5 volumes, but ended up being 12 (numbered I–III and V–XIII), the last one published three years after Hayden had died. The geologically important volumes of the series cover vertebrate fossils (Leidy, 1873; Cope, 1875, 1884), Cretaceous and Tertiary plants (Lesquereux, 1874, 1878, 1883), and Tertiary insects, including a definitive study of the famous fauna from the Florissant beds in Colorado, by Samuel Hubbard Scudder (1837–1911) in 1890. It may be no surprise that none of the volumes is authored by Hayden himself.

Cuvier in France and Richard Owen (1804–1892) in England had advanced vertebrate paleontology to a mature science, but neither of them had available anything like the material the Hayden survey had found in the West, lying around almost on the surface. Like Hall and Barrande before them, Leidy, Cope, and Marsh became the right people in the right places at the right time. Unlike Hall and Barrande, however, all three were operating in the same territory, vast as it was, and that led to conflicts. The differences that arose among them can be understood in the light of that competition.

Cope's contentious character made him difficult to work with, and his strong Quaker outlook made it impossible for him to accept the idea that was then sweeping the world of biological science: Darwin's natural selection. In resisting Darwin (Cope, 1886) he was in good and copious company (with Agassiz, Dana, Dawson, and many lesser lights), and contemporary judgment declared him the loser (by split decision) in his vitriolic war with Marsh, but in present hindsight his paleontology remains sound. He brought order and system to the amazing variety of Cretaceous and Tertiary vertebrate life. Marsh found the toothed birds, and with some help from Thomas Henry Huxley (1825–1895) established the evolution of the horse as a classic Darwinian sequence, but Cope's "Volume III" of the Hayden survey (Cope 1884), with 1009 pages and 134 plates, was an awesome work and is still referred to by vertebrate paleontologists as "Cope's bible."

The publishing program of the Powell survey was less monolithic but even more significant in bringing new ideas to geology. The major himself wrote a partly fictionalized narrative of the exploration of the river (Powell, 1875) and a description of the eastern Uinta Mountains, with an atlas, pointing out the tremendous amount of erosion that had occurred there (Powell, 1876). He also wrote a land-use study covering most of Utah (Powell, 1879).

A report on the geological reconnaissance of the Black Hills, valued mainly for its economic geology, was delayed when Henry Newton (1845–1877) died, but it was published after substantial revision by Gilbert (Newton and Jenney, 1880). The geology of the Henry Mountains (Gilbert, 1877) and the geology of the high plateaus of Utah, with an atlas (Dutton, 1880), appeared under the imprint of the Powell survey, but two of the three most important geological studies conceived in the Powell survey were published as monographs I and II of the new U.S. Geological Survey (Dutton, 1882; Gilbert, 1890). They presented entirely new concepts and major advances in geology.

Two areas of geological understanding were still very hazy in the 1870s: the role of rivers in shaping the landscape and the interpretation of large-scale structures in terms of motions in the Earth's crust. When Major Powell first made his way down the Green and Colorado rivers, he saw the Uinta Mountains, standing huge and inaccessible, and then the Henry Mountains, remote and forbidding. He saw the flexures in the strata of the canyons and it became clear to him that there could be no better place in the world to study these unexplored phenomena. The actual studies were slow getting started, but when Powell finally succeeded in forging his now-famous triumvirate with Dutton and Gilbert, the results poured in.

Powell brimmed with original ideas and knew how to recognize promising lines of research. Credit did not seem to concern him and it is difficult now to decide which particular idea came from which member of the trio. The outstanding monographs of Dutton and Gilbert all warmly acknowledge Powell's help. The encouragement for the work and much of the critical thinking came from him.

Gilbert had risen from unpaid assistant on Newberry's Ohio survey to staff geologist on the Wheeler survey, and he had learned a lot with Wheeler but did not enjoy it. Powell had no trouble enticing him away in 1874. With Wheeler, Gilbert had to follow the line of march. "When the structure of a mountain was in doubt," he writes (1877, p. vii), the geologist "was rarely able to visit the points which should resolve the doubt, but was compelled to turn regretfully away. Not so in the survey of the Henry Mountains." Here he was his own boss and could go wherever he thought he would see the most. Gilbert spent a little more than two months in the Henry Mountains before the first heavy snow drove him out. "Two months would have been far too short a period in which to survey a thousand miles in Pennsylvania or Illinois, but among the Colorado Plateaus it proved sufficient. A few comprehensive views from mountain tops gave the general distribution of the formations, and the remainder of the time was spent in the examination of the localities which best displayed the peculiar features of the structure" (Gilbert, 1877). The way he puts it, one would think it was easy: just hike in, observe, and interpret.

To describe the domed structures produced by large-scale intrusion of magma into sedimentary formations, Gilbert coined the word *laccolite* (later changed to *laccolith*), and he demonstrated that strata which appear rigid on the scale of a hand specimen will bend and deform plastically on a scale of miles. Unlike the Alps and the Appalachians, structures are gentle in the canyon country and the sequence of strata is clearly seen. Powell, Dutton, and Gilbert realized that developing the detailed stratigraphy of the Colorado Plateaus would take decades and they were content to make gross divisions, just enough to allow observation of the major structural features.

Like King, Dutton was also a brilliant writer and parts of his famous monograph (Dutton, 1882) read like a love song to his canyon. Such poetry needed a fitting backdrop and Dutton found just the person to produce it: William H. Holmes (1846–1933), who had already shown his ability as a geological illustrator on the Hayden survey. In the Grand Canyon Holmes found the ultimate subject for his talent. His magnificent panoramas, marvelously accurate in their geology, make the atlas that accompanies Dutton's memoir perhaps the most beautiful American geology book every published.

Dutton's main conclusion was staggering: A geologically very short time is quite enough to allow a river to cut through an enormous thickness of strata rising isostatically in consequence of large-scale denudation. He wrote: " . . . erosion depends for its efficiency principally upon the progressive elevation of a region, and upon its climatal conditions . . . The Colorado River appears to have originated in very early Tertiary time . . . At its beginning its bed lay in Eocene strata . . . The present Grand Cãnon is the work of late Tertiary and Quaternary time" (Dutton, 1882, p. 1, 7–8).

James Hutton had sketched the importance of rivers as agents of erosion, and Lyell had seen their work in Auvergne, but still concluded that coastal wave action was more important, and that opinion was echoed by Darwin. Dana had been impressed with the effect of running water in shaping the volcanic islands of the Pacific, but his view was not widely shared. Not until Joseph Beete Jukes (1811–1869) published his now famous paper on the pattern of river valleys in southern Ireland and its dependence on the structure of the underlying rocks (Jukes, 1862) did the evidence fall into place. That, in retrospect, was the turning point in the fluvialist controversy—a return to Hutton's original idea. Jukes' results were still being actively debated in Britain when the news arrived of what Powell and Dutton had found in the canyon.

FIGURE 13.5. Three directors of two Surveys: C. D. Walcott, J. W. Powell, and Sir Archibald Geikie on a field trip to Harper's Ferry, April 30, 1897. (Photograph courtesy of U.S. Geological Survey; photograph by J. S. Diller)

FIGURE 13.6. Terraces of Lake Bonneville near Wellsville, Utah. (Gilbert, 1890)

The role of spokesman for British geology was then passing from Lyell to another Scot, Archibald Geikie (1835–1924), who became director of the Geological Survey in 1881. In 1879 he had visited Powell and Dutton and, having seen the canyon country, wrote: "Had the birthplace of geology lain on the west side of the Rocky Mountains, this [fluvialist] controversy would never have arisen. The efficacy of denudation, instead of evoking doubt, discussion, or denial, would have been one of the first principles of the science . . . " (Geikie, 1882, p. 183).

Gilbert had first seen the evidence of Utah's ancestral lake when he crossed it with Wheeler. He named it in 1875 after Captain Benjamin Louis Eulalie de Bonneville, U.S. Army (1796–1878), who had explored the Green River area in 1832 and who probably never saw the lake region himself, but who had the good fortune to have his exploits romantically immortalized by Washington Irving (Irving, 1837). Gilbert studied the shorelines of the lake for two years while he was in charge of the U.S. Geological Survey's office in Salt Lake City (1879–1881) and then thought about it for a few years while doing other things.

Geologically and logistically the survey of Lake Bonneville was as different from the survey of the Henry Mountains as it could be. The Henrys were inaccessible, but Gilbert was able to describe much of their geology because he could see so much from the peaks he climbed. Lake Bonneville had covered most of western Utah and had spilled into Idaho. Almost any point in the Great Valley could be reached by wagon, even if it took a long time, but the evidence of the ancient lake had to be collected patiently from a myriad of observations over that hugh area (Gilbert, 1890).

The water level of Lake Bonneville had varied by 1000 feet and produced differential loading of the Earth's crust on a large scale. Gilbert carefully mapped and correlated the old beaches and showed that they are no longer exactly horizontal. The land had bowed up after the lake retreated, and Gilbert was able to estimate the geophysical properties of the crust that would permit the observed amount of uplift. Gilbert's results established geomorphology as a quantitative science and were a revelation in international geology. Dutton and Gilbert had demonstrated that isostatic readjustment in the Earth's crust was the key to vertical movements on a large scale.

CHAPTER FOURTEEN

The Physico-Chemical View

THE MINING INDUSTRY

Western mines had boomed after the Civil War, but it was a thoroughly haphazard boom. Thousands of prospectors combed the mountains, digging here and panning there, without much rhyme or reason. Most of them found nothing of value, and the few who managed to stake promising claims usually lost them to organizers and promoters who were equally ignorant of geology but more familiar with the operations of business. Very few of the mines, including the successful ones, were being designed and operated in a geologically rational manner.

Clarence King was going to change all that. If there is a common thread to his ebulliently varied but ultimately aborted career, it is his conviction that the quantitative, analytical approach is the way in geology and that it would solve all problems. Perhaps it was his old boss Josiah D. Whitney who started him thinking that way in California, where chaos could have been the best single word to describe the mining industry in its early days. As King developed these ideas, he decided that the geologist, with his understanding of structural and stratigraphic relationships, should not only direct the mining operations but also should be able to beat the financiers and business wizards at the mining game. That was one of the reasons why King left the Geological Survey; that was why he continued to invest in the minerals business for the rest of his life. That was where he conspicuously failed in the end. He did not make a fortune for himself, but his frequent calls for rational study of mineral deposits became a guiding force in the spectacular development of the American school of economic geology.

After their first field season in 1867, King's survey set up winter quarters in Virginia City, Nevada, and here they wrote their famous *Mining Industry* (Hague, 1870). It was accompanied by a large atlas of 14 maps and sections, and contained long articles by Emmons, the Hagues, and

King. Two of the papers, both by King, correctly anticipate the tremendous economic potential of two unrelated mining areas, the Comstock Lode for silver, and the Green River Basin for coal. Both of these areas turned out to hold the world's largest deposits of their kind. Wrote King: ". . . it can be said in perfect safety that the series contains a practically inexhaustible supply of coal" (Hague, 1870, p. 457). A paleontological appendix by Meek points out that the coal is of Cretaceous age.

An article on the Toyabe Range, by Emmons, shows his growing insight which later found its best expression in his monograph on Leadville (Emmons, 1886). He is the main author of the concept of contact-metamorphic ore deposition, where mineralizing solutions that originate in igneous intrusions find their way into sedimentary rocks and there deposit metallic ores. He postulated that the water of these solutions had come from the surface and had penetrated into the vicinity of the intruding magma. Later he retreated a bit from that view and ascribed greater importance to "juvenile" water, derived from the magma itself. Almost a century passed before studies of the hydrogen isotopes in rocks and waters showed that his original idea was closer to the truth. Emmons also outlined the processes of secondary enrichment that account for the high local concentration of some metals near the surface.

The Leadville monograph was still in press when two other important economic geologists, both graduates of the Freiberg Mining Academy but a generation apart, joined the U.S. Geological Survey, one bringing the other. The older one was Raphael Pumpelly (1837–1927), probably the best-traveled American geologist up to that time and one of the most versatile. In 1866, after a long sojourn in the Orient, he had been offered the professorship of mining at Harvard and three years later actually moved to Cambridge, but he never spent much time at the university. His interests lay in the real world and he made important studies of the Lake Superior copper deposits. His major works are a study of weathering and the formation of loess (Pumpelly, 1879) and a determination of the structure of the Green Mountains in Vermont and Massachusetts (Pumpelly, 1894). He shares with King the distinction of having recognized the importance of the petrographic microscope and with Charles Richard Van Hise (1857–1918) of having first suggested that the Green Mountains may be an old Alpine structure.

The young man Pumpelly brought with him was a Swede with a passion for the American West. He was Waldemar Lindgren (1860–1939) and he had just finished his studies when he appeared in Washington. He was sent to California, showed his ability there, rose to Chief Geologist in the Survey, and in 1912 moved to MIT to head the Department of Geology. He is best known for his genetic classification of ore deposits

(Lindgren, 1913 and later revisions), presented in a book that remained a standard text for more than two decades.

It would be an exaggeration to say that King, the Hagues, Emmons, Pumpelly, Lindgren, and the rest brought order into the mining business. Important mineral deposits were still being discovered by chance and mined in a haphazard and wasteful manner, but the old way was fading as scientific methods multiplied. In the aftermath of the Silver Panic of 1893, most of the marginal mines had closed and entire mining districts were abandoned. A lot of people suddenly found that their mining stocks were worthless, but the survivors of that shakeout paid attention to new methods.

The U.S. Geological Survey remained the scientific leader in economic geology, but solid research groups were also being established at the University of California in Berkeley, the Colorado School of Mines, the University of Chicago, the University of Wisconsin, Columbia, Yale, Harvard, and a little later, at MIT. Applied geology was in fashion.

Coincidental with that silver panic, the U.S. Geological Survey had a bad year. Major Powell's views on land use and irrigation were, in retrospect, not only liberal, but also scientifically sound. By 1893, however, they had become increasingly unpopular in Congress. Then as now the West was short of water, but the political pressure was for settling and reaping the benefits. The old superstition that "rainfall follows the plow" was still being defended as fact and no one wanted to hear about any limitations imposed by nature. The major was not one to compromise, and as the pressure on him and his Survey mounted, he chose to resign as director to avoid having his old organization cut to shreds by a few vindictive congressmen. He moved over to the Bureau of American Ethnology in the Smithsonian Institution, and in 1894 the Survey began a new chapter with Charles Doolittle Walcott (1850–1927) as director.

Walcott was a fortunate choice. His scientific ability and his special talent for paleontology had been demonstrated long before he was appointed director, but in that post he also became one of the best administrators Washington science had ever seen. He held the job for 13 years and then went on to become Secretary of the Smithsonian.

Walcott started collecting Cambrian fossils when he was still in school; near his home in Trenton Falls, New York, he discovered the first American trilobite with preserved appendages (Walcott, 1876). He never went to college, but worked with James Hall for more than two years, and was able to publish several papers with his own name on them. On Hall's recommendation he was picked by King for the new U.S. Geological Survey in 1879, and there he advanced from lowly assistant to the world's chief Cambrian paleontologist. He was not looking for the oldest fossil then known, but found it in his *Olenellus* zone, now still considered as the base of the Cambrian (Walcott, 1890). Then he beat his own record when he identified unmistakable fossils in the Precambrian Belt series in Montana (Walcott, 1899).

Much later, in 1909, Walcott leading a packhorse up a narrow trail on the barren slope toward Burgess Pass, high above Field, British Columbia (now in the Yoho National Park), saw a block of black shale, split it open, and discovered what is still the best occurrence of soft-bodied marine organisms preserved from Paleozoic time. That was the celebrated Burgess Shale fauna of the middle Cambrian. If he had never done anything else in his life, his work on that alone would have made him a world figure.

Walcott was and always remained very close to Powell, and he understood the major's political troubles. When he took over the reins of the Survey, he skillfully steered it away from political controversies and toward the study of natural resources and basic geology, where he knew the Survey had great strength and powerful support from the mining industry. In three years he had the Survey's gutted budget back in shape and his new program in place; it was not too different from the major's but was geared to broader constituencies. The inner council of the Survey then consisted of Thomas Chrowder Chamberlin (1843–1928), S. F. Emmons, G. K. Gilbert, Arnold Hague, and C. R. Van Hise, and it was an unusual life assemblage—a smooth paleontologist leading a congregation of mappers and hard-rock enthusiasts. It was a golden era for resource geology and for the U.S. Geological Survey.

Among the talented geologists Walcott chose was Florence Bascom (1862–1945) who was mapping major contacts of crystalline rocks from Maine to Maryland. (The first female PhD from Johns Hopkins University, she sat behind a screen in the classroom so she would not distract her fellow students.) She edited the U.S. Geological Survey Philadelphia folio (Bascom, 1909), still the most comprehensive geological description of the area.

GERMAN UNIVERSITIES

The delayed impact of the Industrial Revolution in northern Europe, the rise of the political and military importance of Prussia after the revolutions of 1848, and the ultimate unification of Germany all formed the background for a flowering of science and technology. In the second half of the nineteenth century, German universities became a world force in scientific education, and many young Americans went to Europe to study the sciences. Professors moved freely through the various countries as positions became available in the large German-speaking realm that included not only the German Reich and the Austro-Hungarian Empire, but also Scandinavia, eastern Europe, and Russia. The burgeoning university institutes provided affiliation and employment for young geological talent, just as the surveys were doing in America at that time.

CRYSTALS AND PETROGRAPHY

The main advances in crystallography and petrography came from Europe, particularly Germany. The dominant figures in this intertwined development were Ferdinand Zirkel and Harry Rosenbusch (1836–1914) in petrography, and Evgraf Stepanovich Fyodorov (also transliterated Fedorov or Federov, 1853–1919), Arthur Moritz Schoenflies (1853–1928), Paul Heinrich von Groth (1843–1927), and Victor Goldschmidt (1853–1933) in theoretical and mineralogical crystallography.

Microscopic petrography had been slow to catch on. The art of making thin sections developed in Edinburgh from the optical experiments of David Brewster (1781–1868), but he mentioned it in print only in passing (Brewster, 1814). The polarizing microscope was invented there by William Nicol (1768–1851) and first announced in a two-page article (Nicol, 1829), probably years after the first successful model had been built. Several paleobotanists and mineralogists (including Robert Jameson) studied thin sections of fossil wood and various crystals under the microscope, but the new technique did not create much of a stir even after Henry Clifton Sorby (1826–1908) had demonstrated how useful the polarizing microscope could be in identifying the mineral grains that comprise rocks (Sorby, 1853, 1858).

Young Zirkel visited England in 1860, met Sorby, and they became close friends. The following year Sorby went to Germany and, in his own words, "I strongly impressed on my friend Professor Zirkel the importance of the study of the microscopical structure of rocks; and when sitting on the top of the Drachenfels I described to him how thin sections were prepared, and advised him to apply himself to that kind of investigation" (Sorby, 1870). Evidently Zirkel did not need much coaching, and within a few years he was the world's leading petrographer. As already mentioned, Clarence King soon recognized the importance of his work and arranged for Zirkel to come from Leipzig to New York to select a suite of rocks from the collection made in the survey of the 40th Parallel. More than 2500 thin sections later, Zirkel produced his pioneering *Microscopical Petrography* (Zirkel, 1876), volume VI of King's final report.

The other leading petrographer was Harry Rosenbusch, professor first in Strasbourg and then in Heidelberg. He brought petrography to its full significance and is now remembered for his genetic classification of igneous rocks based on mineralogy as well as chemical composition.

Among the crystallographers, collaboration was truly international. Fyodorov lived mostly in Petrograd (present Leningrad) and managed to obtain an academic position only late in life. His great contribution is the mathematical derivation of the 230 theoretically possible space groups—the fundamental repetitive elements with which space can be completely filled in a symmetrical manner, as in a crystal (Fyodorov, 1891). The full significance of that work did not become apparent until after the discovery of X-ray diffraction by crystals, more than 20 years later, when his space groups became the mathematical basis for the determination of crystal structures. Fyodorov also invented the universal stage for the study of mineral grains in thin section in any orientation.

While Fyodorov was working on his space groups in Petrograd, Schoenflies, a mathematics teacher in the Alsatian town of Colmar, and *Privatdozent* at the University of Goettingen, was approaching the same problem in a different way. Both were in late phases of their research when they finally got together, but both benefited from their correspondence. They published essentially identical results (Fyodorov, 1891; Schoenflies, 1891) and unwittingly created a lot of work for historians. Ever since, many Russian writers have felt obliged to defend Fyodorov's priority, while German commentators made it clear that the Schoenflies derivation is more general and purer mathematically, without any implication of physical constraints. The whole situation is made even more complicated by the studies of William Barlow (1845–1934), an English gentleman-scientist, who also worked out the 230 space groups independently and in still another way (Barlow, 1894).

The optical goniometer for measuring the interfacial angles of small crystals by reflecting a thin beam of light from them had been designed by Wollaston, mainly to check the geometrical laws of crystal morphology that had been proposed by Haüy. The original instrument (Wollaston, 1809) was capable of a precision of a minute of arc and subsequent improvements by many other workers ultimately increased the precision to a few seconds. Victor Goldschmidt, who like Sorby and Barlow was financially independent and sought no academic post, was the great compiler and systematizer of the data on crystal morphologies obtained with these instruments. He had graduated under Rosenbusch in Heidelberg and for many years chose to work in his institute there, but the two men had little in common and never worked together on any major project.

Von Groth was professor at Strasbourg and Munich, and he became the main factor in the early understanding of the physical significance of the crystal lattice and the relation of chemical composition to crystal structure. As founder and editor of the *Zeitschrift für Kristallographie*, he acted as coordinator and catalyst of the rising science of crystallography.

THE EARTH'S CORE

The most important geophysical development of the second half of the nineteenth century was the growth of seismology. The term was coined by Robert Mallet (1810–1881), an imaginative and incredibly prolific engineer whose interest in earthquakes was first inspired by the writings of Charles Lyell. He worked on diverse engineering projects, from the casting of heavy ordnance to the corrosion of iron and steel by sea water, but always returned to seismology. He compiled the first modern catalog of recorded earthquakes and the first seismic map of the Earth (Mallet, 1849–1871, 1850–1858), built several seismographs, and measured seismic velocities in the field. Another English engineer, John Milne (1850–1913), in 1875 found himself appointed Professor of Mining and Geology at the Imperial College of Engineering in Tokyo. There, as he later put it, he "had the opportunity to record an earthquake every week" (Milne, 1886, p. vii).

Beginning with the pendulum principle, Milne and his colleague at the Imperial College, Thomas Gray (1850–1908), developed a horizontal-motion seismograph that recorded mechanically on a sheet of paper blackened with soot from an oil lamp and wrapped around a slowly rotating clockwork-driven drum (Milne, 1886, p. 39). The device was sensitive enough to record distant earthquakes and portable enough to be taken to far-flung places.

A group of British engineers and scientists then working in Japan, together with their Japanese counterparts, all spearheaded by Milne, established the Seismological Society of Japan, enlisted the assistance of the government in transmitting time signals by telegraph, and set up the first seismograph network. Milne had developed methods for locating distant earthquakes and actively compiled the data for establishing travel-time curves for the various waves. He was primarily responsible for the initial operation of the system presently used for gathering and publishing worldwide seismological data.

About 1900, Emil Wiechert (1861–1928), Professor of Geophysics at Goettingen, introduced a new seismograph of greatly improved sensitivity. It was an astatic instrument, based on the principle of the inverted pendulum, and consisted of a very large mass (a ton or more), pivoted underneath and held in equilibrium position by light springs. With the same old mechanical recording, the Wiechert pendulum gave a faithful record of actual ground motion, which permitted close analysis and the identification of the arrivals of the various waves reflected and refracted by discontinuities within the Earth (Wiechert, 1908).

Ever since Maskelyne's and Cavendish's determinations of the mean density of the Earth it had been clear that the Earth must have a heavy interior. Many had postulated a dense core, but there was no way of determining its size until worldwide earthquake seismology offered the means. Richard Dixon Oldham (1858–1936) first showed the existence of a core from seismic observations (Oldham, 1906), but his estimate of the diameter (about 5000 km) was too low because he had misinterpreted the evidence from distant earthquakes. Wiechert's student Beno Gutenberg (1889–1960) made the first correct interpretation of the arrivals of seismic waves reflected and refracted by the core, and calculated its diameter as about 7000 km, a figure that still stands (Gutenberg, 1913).

CHAPTER FIFTEEN

Cooling Earth, Flexing Continents, Radioactivity

HEAT OF THE SUN AND THE EARTH

The greatest personification of Victorian rationalism in science was surely William Thomson (1824–1907), Professor of Physics at the University of Glasgow, who was elevated to Lord Kelvin and later Baron Kelvin of Largs. In terms of permanent scientific value Kelvin's most important work was in thermodynamics, but he was also interested in many other things and became famous as the scientific hero of the transatlantic cable when he diagnosed its early failure as having come from excessive current loading. He remedied the problem by replacing the crude receiving relays with sensitive galvanometers he had designed.

From his early student days and throughout his long career, Kelvin was fascinated by the energy balance of the Earth and the sun, and that brought him into contact (and ultimately into conflict) with geologists. The origin of the sun's heat was a problem and by the middle of the nineteenth century it was clear to physicists that the sun could not remain hot forever. Its thermal energy was very large, but in the absence of any conceivable replenishing process it had to be decreasing.

Kelvin had not been the first to think about this, but he was easily the most dynamic expositor of the new idea. He added up the amounts of energy emitted by the sun, assumed physical conditions and heat capacities that were plausible at the time, and calculated that the sun was cooling quite rapidly—on a geological time scale (Thomson, 1862*a*). Every movement of air and water on the surface of the Earth is produced by the energy received from the sun. Weathering of rocks and transport of sediments—every kind of erosion—ultimately derive from the energy of the sun. What was the meaning of uniformitarianism if the sun was cooling? All geologic surface processes must be slowing down.

Kelvin's early geological ideas were influenced by his American friend and senior colleague at the university, Henry Darwin Rogers. In Virginia, Henry's brother William had observed high temperatures in coal mines, and Henry very likely discussed these findings with Kelvin, who was keenly interested in heat conduction in the Earth. His first published papers had been on that subject.

For years he worked on the problem, obtained data on temperatures at various depths, made proper assumptions about the heat conductivity of the intervening rock masses, and estimated the outward heat flow. He concluded that the time since a crust first formed on a presumably molten globe was about the same as the age of the sun—perhaps 100 million years (Thomson, 1862*b*). Like Descartes before him, Kelvin also had no idea of what the temperature of the Earth might have been when it first formed, but in these calculations the figure makes practically no difference (Bullard, 1976) because the cooling of an entirely molten globe is extremely rapid compared with one that has a solid crust. Considering all the assumptions that Kelvin had to make in his calculations, he chose to ascribe wide limits of uncertainty to his result: a factor of four. The Earth and the sun could have been anywhere from 25 to 400 million years old.

In the beginning, geologists paid no attention to Kelvin's attacks on uniformitarianism, and that annoyed him. Being ignored was one thing he disliked, so he decided to jolt them by attacking on their home turf. In 1868 he read a paper on geologic time to the Glasgow Geological Society, and now they took notice. In response, Thomas Huxley spoke up for uniformitarianism (Huxley, 1869) in a speech that was smooth and elegant, as usual, but carried little force because its arguments brought up nothing that Kelvin had not already considered. Then as now, it was difficult for geologists to argue with the results of physics.

Kelvin's approach was direct, his premises made sense, and the uncertainties he quoted were large enough to allow most geologists to make an orderly retreat from the concept of very long geologic time. As the years went by, Kelvin's calculations gradually tightened and his time became shorter and shorter. In 1880, Clarence King and Carl Barus (1856–1935) set up a laboratory in the U.S. Geological Survey to measure physical properties of rocks, including thermal conductivity and heat capacity (Barus, 1893). On the basis of the new data King greatly reduced the uncertainty of Kelvin's calculations of the cooling rate. Most geologists had already capitulated to the overwhelming power of Kelvin's physics when King presented his new age for the Earth's crust: 24 million years (King, 1893).

King's time span was overwhelmingly logical, impeccably founded in contemporary physics—and completely wrong, as we now know. In a way it epitomized the condition of geology at the time: confidence in new analytical techniques, rapid growth of reliable data, and no clear suggestion of where it was all going. Kelvin's time could not accommodate the data of stratigraphy and organic evolution; the widely observed shortening of the Earth's crust could not be physically explained; the sea floor was too small to receive all the sediments produced in denudation of the land; the forces of mountain building evaded understanding.

GEOTECTONICS

In Europe, the most influential geologist of the time was Eduard Suess (1831–1914). Born in London of German parents, he grew up in Vienna, studied there and in Prague, never took a degree but became an accomplished linguist, and ultimately became Professor of Paleontology at the University of Vienna. (He also had a taste for politics, served on the city council and in Parliament, and became rector of the university.)

Suess worked in many fields—from graptolites to volcanism to sanitary engineering—but he is remembered for his conceptualization of structural geology. If the thrust of Lyell's time was toward stratigraphy, the time had come now for a new direction: tectonics. Suess' work on the origin of the Alps (Suess, 1875) was an analysis of mountain-building processes quite beyond the Alps themselves. He formulated the concept of the Tethys, the large ancestral Mediterranean Sea, and recognized the compressive nature and fundamental asymmetry of Alpine chains.

His literary masterpiece is a monumental synthesis of the Earth's structure, the five-volume *Face of the Earth* (Suess, 1883–1909; English translation, 1904–1924; French translation, 1897–1918). It became the geologists' bible in his time. Suess' approach was comparative, open-minded, and encyclopedic. He had a broad command of the geological literature in many languages and used it to give substance to his syntheses. As he was writing, he often found new evidence and did not hesitate to contradict the conclusions he may have reached earlier.

Suess viewed the ancient continental shields as stable, and the oceans as major areas of collapse. The bottom of his ocean determined the level of the sea, and ocean-bottom subsidence produced the "eustatic" sea level changes he was observing. All water came from the outgassing of the planet's interior, and magmatic processes still contribute "juvenile" water to the "vadose" water now on the surface. His marine subsidence was a contradiction of the shrinking-Earth concept, and he was aware of that, but the geosynclinal mountain building in the margins of the Pacific and the Tethys required it, and he was willing to leave the resolution of that problem to his successors.

Suess' global tectonic ideas were developed further by Wilhelm Hans

Stille (1876–1966), a Hanoverian who graduated from Goettingen in 1899, worked for the Prussian geological survey, taught at the universities of Hannover, Leipzig, Goettingen, and Berlin, and developed a following even wider than Suess'. After about 1930, Stille's students and followers held a virtual monopoly on the teaching of tectonics in Germany.

Stille saw the world as a system of stable and fixed "cratons" (a word he coined) separated by geosynclinal belts. He developed the connection of large-scale plutonic intrusion and volcanism with geosynclinal environments, and correlated these events into a worldwide sequence of orogenic pulses. In all that work, stratigraphy was his only yardstick of time, of course. Stille did live to see the development of nuclear age determination, but by that time his work was done.

The work of Suess and Stille meshed widely with contemporary geology. Their tectonic syntheses were based on data that came to them from all over the world, and Suess' many disciples further enlarged "the big picture." Prominent among them were Marcel Bertrand (1847–1907), Pierre Termier (1859–1930), Émile Argand (1879–1940), F. A. Vening Meinesz (1887–1966), Walter Bucher (1889–1965), and Marshall Kay (1904–1975). As they strove to understand and generalize the deformations of the Earth's crust, they mainly succeeded in making its character more complicated. To account for the large-scale compressive effects observed all over the world, they had to assume that the Earth's crust was shrinking, but no one was able to come up with a viable process to bring about the necessary diminution. All attempts to understand the tectonic history of the Earth became hopelessly snagged on that basic obstacle. Argand, with his extraordinary imagination, was the only one among them to recognize early that continental drift may offer clues to the problem.

The dilemma continued while the two ideas that were destined to resolve it appeared and were being developed slowly and somewhat inconspicuously by small bands of enthusiasts. For decades, the bulk of the geological profession took little notice of these developments. There was no crucial meeting, no sudden turn of events, no "revolution." For very good reasons it took a long time before the two ideas finally could have their impact on the mainstream of geological thinking. The first of the two ideas was radioactivity and the second was continental drift.

RADIOACTIVITY

When Henri Becquerel (1852–1908) was appointed to the prestigious chair of physics in the École Polytechnique in Paris in 1895 it seemed obvious to many that he was past the peak of his distinguished research career, but as Alfred Romer observed much later (1970, p. 558), "everything for which he is now remembered [was] still undone." Barely ensconced in his new laboratory in the rooms at the *Jardin* where Cuvier once lived, Becquerel learned of Wilhelm Conrad Roentgen's (1845–1923) discovery of a new radiation that was produced when "cathode rays" (now known to be accelerated electrons) were made to impinge on a screen in a discharge tube. Long ago Becquerel had worked on the luminescence of crystals with his famous father, and Roentgen's radiation intrigued him. He thought it might be some kind of phosphorescence, and within a few weeks he had found that crystals of potassium uranyl sulfate, which were known to fluoresce in sunlight, also emitted a penetrating radiation like Roentgen's, which blackened a photographic plate even when it was wrapped in black paper. By May 1896, he knew that this was not phosphorescence, but that he had found an inherent new property of uranium (Becquerel, 1896).

From there it was a long way to the discovery of radioactive decay, but many were looking for explanations of the puzzling radiation. The first glimmer of radioactive transmutation came from the emanation experiments of Elster and Geitel (1902) in Wolfenbüttel, and Rutherford and Soddy (1902) in Montreal. Ernest Rutherford (1871–1937) wrote the basic equations of radioactive decay in 1904, calculated the amount of heat released by the radium that is present in ordinary rocks, and concluded that "the time during which the Earth has been at a temperature capable of supporting the presence of animal and vegetable life may be very much longer than the estimate made by Lord Kelvin from other data" (Rutherford, 1904, p. 346). Shortly afterward, Robert John Strutt, later Lord Rayleigh (1842–1919), showed that the amounts of helium accumulated in minerals from the decay of uranium are much larger than could have been produced in Kelvin's geologic time (Strutt, 1905).

At Yale University, Bertram Boltwood (1870–1927) collected published chemical analyses of pure uranium minerals (mostly made by Hillebrand, 1890) and calculated their ages from their lead content (Boltwood, 1907). His ages ranged up to 2000 million years, and some of them are not far from present values because errors in his constants tended to compensate. Whatever their inaccuracy, Boltwood's results clearly reestablished a long geologic time scale.

The theoretical foundations for the measurement of geologic time by analysis of the products of radioactive decay were laid, but the technology required for actually making useful measurements was still far away. Gravimetric chemistry was woefully inadequate for determining the small amounts involved, the systematics of radioactive decay chains (of uranium and thorium) were only beginning to be understood, and the rates of radioactive decay were known either roughly or not at all. It was more than 50 years after Becquerel's discovery that age measurements by nuclear methods became available to geologists in sufficient numbers to be of much use.

That period coincides almost exactly with the long career of Arthur Holmes (1890–1965). He began in Strutt's laboratory in London in 1910, working on uranium and lead, and his first paper presaged his future work on determinations of geologic time (Holmes, 1911). Knowledge of decay schemes and constants had greatly improved in those few years since Boltwood's work, and Holmes' results were much more reliable, but they generally confirmed the earlier work. Holmes' main problem was geological: the placement of the unusual uranium mineral occurrences he was then able to date in the stratigraphic column. He proposed the first of his famous time scales in a popular booklet rather than a scientific publication (Holmes, 1927), perhaps because he knew that it was based more on his geologic intuition than on hard physical and stratigraphic data. His subsequent revisions (Holmes, 1937, 1947) continued in a light vein, but analytical technology caught up with theory in the 1950s.

Arthur Holmes' First Published Time Scale—Comparing Ages Based on a 30 Million Year Cycle with Ages from Analyses of Uranium Minerals

		Ages in Millions of Years		
		By Adding Successive Cycles	From Radio-Active Minerals	
0.	Present Day	0		
1.	Upper Tertiary	30	35	
2.	Lower Tertiary	60	60	
3.	Upper Cretaceous	90		
4.	Lower Cretaceous	120		
5.	Jurassic	150		
6.	Triassic	180		
7.	Permo-Triassic	210	205	
8.	Permo-Carboniferous	240	220	
9.	Upper-Carboniferous	270	240	
10.	Lower Carboniferous (*b*)	300		290
11.	Lower Carboniferous (*a*)	330		375
12.	Upper Devonian	360	300	
13.	Middle Devonian	390		
14.	Lower Devonian	420		
15.	Silurian	450		

16.	Upper Ordovician	-	480	440(?)
17.	Middle Ordovician	-	510	
18.	Lower Ordovician	-	540	
19.	Upper Cambrian	-	570	
20.	Lower Cambrian	-	600	
	Late pre-Cambrian	-		575 590 640

(Adapted from Holmes, *The Age of the Earth*, 1927; courtesy Ernest Benn Limited.)

Isotopic age determinations on rocks securely tied to the fossil record became available. Bracketed intrusives, igneous rocks that intruded sedimentary rocks whose relative age is known from their fossils, were reported (Faul, 1959, 1960). Holmes developed his justly famous and remarkably accurate "1959 scale" (Holmes, 1960) using the new data. It is still the yardstick used in the study of geologic time.

The Manhattan Project (1942–1946) jolted nuclear analytical technology, including refinement of mass spectrometry developed so brilliantly by Alfred O. Nier (1911–) just before the war (Nier, 1939). After his development of the technique of the calculation of geologic time from the decay of uranium to lead, the techniques of argon accumulated from the decay of potassium and strontium from the decay of rubidium became tools in the arsenal of determining geologic time. These analytical methods all measured long periods of time in the Earth's history. Dating younger material by the radiocarbon method was developed by J. R. Arnold and Willard F. Libby (1949). Libby (1908–1981) refined the technique.

Refinement of methods in determining the age of rocks was combined with other evolving geologic studies: paleomagnetism, the realization that great ridges ran through the middle of all the Earth's oceans, and magnetic profiles measured by ships over these ridges. The sea floor was shown to be spreading, proving a hypothesis that had emerged again and again throughout the history of geology—continental drift.

CHAPTER SIXTEEN

Continental Drift, Subsiding Ocean Basins, Things to Come

CONTINENTAL DRIFT

The tetrahedrally symmetrical position of the continental masses on the Earth's sphere was noticed as world maps became reasonably accurate. Francis Bacon (1561–1626) in *Novum Organum* in 1620 pointed out the general conformity of the outlines of Africa and South America. He did not actually compare the outlines of the continental coasts on both sides of the Atlantic Ocean, however. Atlantis, the mythical land mass mentioned throughout history, entered the geological literature. Bacon suggested that remnants of it may have formed North and South America. Father François Placet proposed, quite poetically, in the booklet *La Corruption du Grand et Petit Monde* published in Paris in 1868, that before the Noachian flood, the land must have been continuous: The Deluge caused the separation of the land masses by uplift and subsidence. The Comte de Buffon in the late eighteenth century speculated about a former land of Atlantis in the present Atlantic Ocean. Buffon cited paleontological evidence, the similarity of fossils in Ireland and America, as proof of this land.

The possibility that the continental masses might have drifted away from each other was mentioned in an obscurely-written book by Richard Owen (1810–1890), long-time professor at the University of Indiana and brother of David Dale Owen (R. Owen, 1857). It was to be part of a grand and slightly mystical scheme of nature, and it received a fair amount of discussion at the time, but was soon forgotten.

In 1858, Antonio Snider, an American living in Paris, described in *La*

Creation et ses Mysteres Devoiles the breakup and drifting apart of the Atlantic continents. An old-fashioned catastrophist, he speculated that during the Earth's formation from the cooling of a molten mass, the continents had formed on one side only, thus creating a condition of instability. The Noachian flood caused enormous fracturing and splitting of the land and the island of Atlantis drifted westward, finally becoming the Americas. His compelling evidence of the split was the coastal fit of Africa and South America. His descriptions of uniting and splitting continents were later echoed in Taylor and Wegener's hypotheses early in the twentieth century.

The consideration of continental drift continued to be associated with catastrophic events. Reverend Osmond Fisher (1817–1914), rector of Harleton, England, postulated in 1881 in *Physics of the Earth's Crust*, perhaps the first textbook in geophysics, that the Earth's relatively fluid interior was subjected to convection currents rising beneath the oceans and falling beneath the continents. He proposed in the journal *Nature*, also in 1881, the catastrophic event that the moon had been wrenched from the Earth with the resulting scar of the Pacific basin. Molten lava at depth began to fill this cavity and the light granitic crust above broke into fragments that drifted slowly toward the depression. His ideas were largely ignored.

Half a century after Richard Owen suggested that continental masses drifted from each other, Frank Bursley Taylor (1860–1938) of the U.S. Geological Survey came forth with a much clearer but at the time no more plausible suggestion that movements of continents are the cause of Tertiary mountain building on a global scale (Taylor, 1908, 1910). He noticed the position of the mid-ocean ridge, halfway between Europe and America, and thought it was reasonable to postulate "that the mid-Atlantic ridge has remained unmoved while the two continents on opposite sides of it have crept away in nearly parallel and opposite directions" (Taylor, 1910). Taylor was a respected Pleistocene geologist, but his global tectonic idea, almost prophetic as it was, caused almost no discussion.

Anglo-American geophysical tradition laid emphasis on the properties of the solid Earth whereas in Germany scientists integrated meteorological, climatological, and geophysical data and adopted mobilist views—segments of the Earth's crust floating on a liquid interior. Alfred Wegener (1880–1930) surely was influenced in the development of his theories by his countrymen. The concept of continental drift was presented "not for the first time perhaps, but for the first time boldly" (Marvin, 1973, p. 85) by Wegener, German astronomer, geophysicist, and meteorologist—but above all an adventurer. [He loved skiing and mountain climbing and, in 1906, together with his brother Kurt, set a world endurance record in free ballooning (Marvin, 1973).]

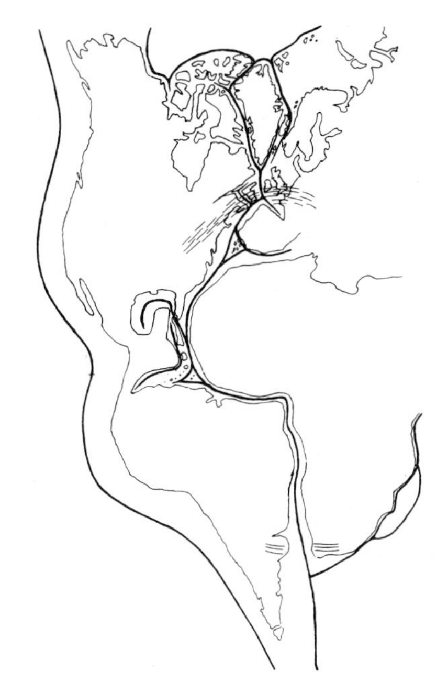

FIGURE 16.1. Wegener's reconstruction of the pre-Atlantic continental blocks. (Wegener, 1915; courtesy of Friedr. Vieweg & Son)

In constructing his theory, Wegener began with the well-known similarity of the South Atlantic coastlines and expanded that observation through tectonics, paleontology, and paleoclimatology into a daring world picture of a huge primordial continent breaking up and the fragments drifting away from each other to form the present configuration of land and sea.

From the beginning Wegener insisted in print that the thought of continents drifting came to him independently, but he also made it clear that he was a thorough and imaginative searcher in libraries. He had read Taylor's work and acknowledged it (Wegener, 1912, p. 185). He may well have known about it in 1911 or even a bit earlier (Totten, 1980)—at least a year before he himself had said anything on the subject in public.

He first outlined his ideas in two lectures in Frankfurt and in Marburg in January 1912. He was about to leave on his second expedition to the Greenland ice cap, and aware that he might not come back, he carefully arranged to have the subject of the lectures published that year (Wegener, 1912). A year later he did come back, just in time to be caught up in World War I. He was wounded and while recuperating at home, he returned to the subject of drifting continents and expanded his earlier paper into a slender paperback, *Die Entstehung der Kontinente und Ozeane*, which was printed on poor paper and apparently in a small edition (Wegener, 1915). Probably because of the war, it went almost unnoticed.

When the war was over, he brought out two more editions (Wegener, 1920, 1922) before the Allied world finally overcame its distaste for things German, and the third edition was translated into English (Wegener, 1924), French, Spanish, Swedish, and Russian. In the 1920 edition Wegener fit all the continental blocks on a globe as he determined their positions in the Permocarboniferous, and in the 1922 third edition he published his global depictions of a Carboniferous protocontinent, Pangaea, to its breakup in the early Quaternary. This time the spark was struck and a vigorous debate ensued, first in England (Lake, 1922, 1923) and later in America (Van der Gracht, 1928). It lasted a little more than a decade, but the geologists were not ready for Wegener.

Wegener's evidence was powerful but not sufficient and that one fatal obstacle—the absence of any plausible force that could move the continents, soon made ever so clear by the elegant geophysical mathematics of Harold Jeffreys (1891—) (Jeffreys, 1924, 1929) and still a subject fit for debate a half-century later—kept the geological establishment from coming aboard. Émile Argand in France and Arthur Holmes in England gave qualified support to the drift idea and Alexander du Toit (1878–1948) in South Africa broadcast it with vigor (du Toit, 1937), but the bulk of informed geological thinking on the subject rarely went beyond treating continental drift as an intriguing but "purely speculative idea" (Holmes, 1944, p. 508).

Wegener himself took the position that continental drift had been proven conclusively by the geodetic measurements at Kornok in Greenland, which showed to his satisfaction that Greenland had drifted 180 meters westward in five years from 1922 to 1927. In the fourth edition of his book (Wegener, 1929), he suggested that geologists get over debating the existence of continental drift and get on with investigating its consequences. He had conspicuously ignored various bits of evidence that did not fit his scheme, however, and his confidence in the accuracy of the geodetic determinations at Kornok was misplaced. His credibility questioned, he was making no headway convincing the world.

His third expedition to Greenland was the most ambitious. He had raised the money himself and this time he was in command, but things did not go well. The first phases of the project, in 1929, went smoothly but in the spring of 1930 the sea ice was heavy and they were more than a month late getting off their ship and onto the ice north of Umanak on the west coast. They learned their new propeller-driven sleds used so much gasoline that they could not have made it across the ice cap with a useful load, and they had to be abandoned. That meant that radio equipment and a lot of geophysical gear, including the seismic rig they brought to explore the thickness of the ice cap, as well as the small portable building they intended as winter quarters for the station they had set up midway on the ice, all had to be left behind on the coast (Schwarzbach, 1980).

A less-adventurous commander would have scrubbed the mission right there, but Wegener was in a position where he had to go on. With two men and dogsleds carrying insufficient supplies and a start too late in the season, he headed east to resupply the midway station at 71 degrees north latitude where two men were holed up in a tent. He expected the trip to take no more than 20 days but it took 40. Having used up most of the food and fuel they were carrying, Wegener spent only one day, his fiftieth birthday, before he headed west again. The two men at the station had barely enough supplies to last the winter, but one of his trip companions had severe frostbite and had to be left with them.

It was now November and the polar night covered the ice with semidarkness while temperatures dropped to merciless levels. Skiing behind their dogsleds, Wegener and his remaining partner Rasmus Villumsen, a sturdy young Eskimo, traveled westward but never reached the coast. Next spring a search party found Wegener's body almost exactly halfway down the trail, neatly buried in his sleeping bag, between his skis stuck upright in the firn. The dogs, the sleds, and Villumsen were never found.

AFTER WEGENER

After Wegener, geophysicists, especially, jumped into the abyss of continental drift theories. Beno Gutenberg in Germany (Gutenberg, 1927), determined from seismic evidence that the moon had been torn from the Pacific floor and that especially those land masses opposite the scar had spread, causing mountain ranges and shaping continents. Arthur Holmes' absolute time-scale, based on the decay rates of radioactive elements in rocks, supported Wegener's evidence for continental drift. Once convinced, he became a lifelong supporter in print from 1928 onward. He reexamined an old idea of geophysics that the interior of the Earth is in a state of extremely sluggish thermal convection—trenches mark the places where currents in the mantle descend again into the interior of the Earth, pulling down the ocean floor. The South African, Alexander du Toit, became an ardent—one might say dramatic—supporter of Wegener and continental drift in *Our Wandering Continents* (du Toit, 1937). He introduced modifications such as the separation of Wegener's Pangaea into supercontinents, the northern one called Laurasia and the southern one Gondwanaland. His support was a counter to the ridicule emerging from the nonbelievers, especially in America.

By the late 1940s, the study of paleomagnetism, which indicates in certain kinds of old rocks the direction of former magnetic fields of the Earth, was sophisticated enough to indicate either migration of the Earth's poles or migration of the continents. Edward Bullard, S. K. Runcorn, and others of a group at Cambridge University recognized that paleomagnetism determinations indicated that Taylor's and Wegener's hypotheses could be explained by closing together the North American and European continents. From about 1956 onward, more and more geologists converted to the concept of continents moving relative to the magnetic poles and to one another.

Maurice Ewing of Columbia University and Bruce C. Heezen of Lamont Geological Observatory gathered enough evidence to propose that a system of mid-ocean ridges extended continuously for 40,000 miles through all the world's oceans. Measurements first undertaken by Bullard showed that the flow of heat is unusually great along these ridges. In 1960, Harry H. Hess of Princeton University suggested that the sea floors crack open along the crest of the mid-ocean ridges and that new sea floor forms there and spreads apart on either side of the crest. Wegener had held that each continent is independently propelled; Hess proposed that the continents move as "rafts" frozen into and moving with a rigid sea floor. Robert S. Dietz named this process of new floor generated at mid-ocean ridges "sea-floor spreading"—integral with that interpretation was the hypothesis that where continents and sea floor come together, subduction zones occur where the sea floor is absorbed beneath zones of deep ocean trenches and young mountains.

A landmark in the history of the earth sciences in the twentieth century was reached in 1963 by a profound interpretation of the germinal ideas first advocated so forcefully by Wegener. Fred J. Vine and D. H. Matthews, then at Cambridge University, proposed that the hypothesis of the continuous creation of ocean floors might be tested by examining the magnetic pattern on both sides of an oceanic ridge. The extraordinary correlations discovered by them with their zebra-stripe charts and by other workers are illustrated and described in the geological literature of today. Continental drift, sea floor spreading, plate tectonics are the topics for future historians of geology.

REFERENCES

Acta (1780). Acta Sacrae Facultatis Theologiae Parisiensis Occasione Libri qui inscribitur: *Histoire Naturelle, générale & particulière, contenant les Époques de la Nature, Paris 1778*, (Paris, 1780).

Adam, Charles and Paul Tannery (1971). *Meditations . . . Principes*, B. Rochot, ed., Vol. 9 (2), Oeuvres de Descartes, Nouvelle présentation en co-edition avec le Centre National de la Recherche Scientifique (1897–1913), Paris.

Agassiz, Louis (1841). *Untersuchungen über die Gletscher* (or *Études sur les Glaciers*), Jent & Gassmann, Solothurn, Switzerland.

Agassiz, Louis (1833–1844). *Recherches sur les poissons fossilis*, 5 vols., Neuchâtel.

Agassiz, Louis and Elizabeth Cabot Cary Agassiz (1868). *A Journey in Brazil*, Ticknor & Fields, Boston.

Agricola, Georgius (1530). *Bermannus sive de re metallica dialogus*, Froben, Basel.

Agricola, Georgius (1546). *De ortu & causis subterraneorum*, lib. V. *De natura eorum quae effluunt e terra*, lib. IV. *De natura fossilium*, lib. X. *De ueteris et novis metallis*, lib. II. *Bermannus sive de re metallica dialogus liber. Interpretatio Germanica uocum rei metallicae*, Froben, Basel.

Agricola, Georgius (1556). *De re metallica*, lib XII. Froben, Basel.

Agricola, Georgius (1557). *Vom Bergwerck Bücher*, Philipus Bechius, trans., Froben, Basel.

Agricola, Georgius (1558). Extensively revised edition of Agricola, 1546, Froben, Basel.

Agricola, Georgius (1912). De re metallica, Herbert Clark Hoover and Lou Henry Hoover, trans. & ann., *Mining Magazine*, London; reprinted Dover, N.Y. (1950).

Albertus Magnus (1476). *De mineribvs et rebvs metallicvs*, 1st ed. Padua.

Albritton, Claude C. (1980). *The Abyss of Time*, Freeman, San Francisco.

Aldrovandi, Ulisse (1648). *Musaeum metallicvm*, in libros III distributum, Ferronius, Bologna.

Arbuthnot, John (1697). *An Examination of Dr. Woodward's Account of the Deluge . . .*, printed for C. Bateman, London.

[Aristotle] (1936). *Aristotle Minor Works*, W. S. Hett, trans., Harvard, Cambridge (Mass.).

Aristotle (1952). *Meteorologica*, H. D. P. Lee, trans., Harvard, Cambridge (Mass.).

Arnold, J. R. and Willard F. Libby (1949). Age Determinations by Radiocarbon Content: Checks With Samples of Known Age, *Science*, Vol. 110, no. 2869, pp. 678–680.

Association of American Geologists and Naturalists (1843). *Reports of the First, Second, and Third Meetings . . .* , Gould, Kendall & Lincoln, Boston.

Atwater, Couleb (1820). Letter to Benjamin Silliman, *Amer. Jour. Science*, Vol. 2, pp. 242–246.

Aufrère, L. (1952). De Thalès à Davis. *Le relief et la sculpture de la terre (auteurs, textes, doctrines, ambiences,)* Vol. IV. La fin du XVIIIe siècle. I. Soulavie et son secret. *Un conflit entre l'actualisme et le créationisme du temps géomorphologique*, Hermann, Paris. 193p.

Avicenna (Ibn Sīnā) (1021–1023). *De mineralibus et rebus metallicis* in *The Book of the Remedy.*

Azophi (Al-Sufi). *Book on the Constellations*, see T. Hyde.

Babington, William (1799). *A New System of Mineralogy . . .* , Phillips, Robinson, Cox, London.

Bache, Alexander Dallas (1856). Notice of Earthquake Waves on the Western Coast of the United States, on the 23d and 25th of December, 1854, *Amer. Jour. Sci.*, Vol. 21, pp. 37–43.

Bache, Alexander Dallas (1863). *Records and Results of a Magnetic Survey . . .* , Smithsonian Institution, Washington.

Bacon, Francis (1620). *Novum organum*, G. Kitchin, trans. (1855) Univ. Press, Oxford.

Baier, Joh. Jacob (1708). *Oryctographia norica . . .* , Michahellis, Nuremberg.

Baier, Joh. Jacob (1958). *Johann Jacob Baier's Oryctographia Norica nebst Supplementen*, Bruno V. Freyberg, ed., Hermann Hornung, trans., Erlanger geologische Abhandlungen Heft 29, Erlangen.

Bakewell, Robert (1813). *An Introduction to Geology . . .* , Harding, London.

Barlow, William (1894). Ueber die geometrischen Eigenschaften homogener starrer Structuren und ihre Anwendung auf Krystalle, *Zeitschrift für Krystallographie*, Vol. 23, pp. 1–63.

Barrande, Joachim (1846). *Notice préliminaire sur le Système Silurien et les trilobites de Bohême*, Leipzig, pp. 6, 97.

Bartlett, Richard A. (1962). *Great Surveys of the American West*, University of Oklahoma Press, Norman.

Barus, Carl (1893). High-Temperature Work in Igneous Fusion and Ebullition Chiefly in Relation to Pressure, *U.S. Geol. Survey Bull.*, no. 103, 57 p.

Barus, Carl (1893). The Fusion-constants of Igneous Rock, Part 3, the Thermal Capacity of Igneous Rock . . . , *Phil. Mag.*, Vol. 35, pp. 296–307.

Bascom, Florence (1909). Philadelphia Folio, *U.S. Geol. Surv. Geol. Atlas*, no. 162.

Bass, George F. (1967). Cape Gelidonya: A Bronze Age Shipwreck, *Trans Amer. Phil. Soc.*, Vol. 57 (ns), pt. 8, 177 p.

Beaumont, John, Jun. (1693). *Considerations On a Book, Entituled The Theory of the Earth. Publisht some Years since by the Learned Dr. Burnet*, Taylor, London.

Becquerel, Henri (1896). Sur les radiations invisibles émises par les corps phosphorescents, *Comptes Rendus*, t. 122, pp. 501–503.

Beringer, Joannis Bartolomaeus Adam (1726). *Lithographiae Wirceburgensis . . .* , Fuggart, Würzburg.

Beringer, Joannis Bartolomaeus Adam (1963). *The Lying Stones of Dr. Johann Bartholomew Adam Beringer, being his Lithographiae Wirceburgensis*, Melvin E. Jahn and Daniel J. Woolf, trans. & ann., Univ. of California Press, Berkeley.

Biringuccio, Vanoccio (1540). *De la pirotechnia*, Venice.

Biringuccio, Vanoccio (1943). *The Pirotechnia of Vanoccio Biringuccio*, Cyril Stanley Smith & Martha Teach Gnudi, trans. & ann., Amer. Inst. Mining Metallurg. Eng., New York.

al-Biruni (1910). *Al-Beruni's India*, E. Sachau, trans. & ed., Trubner, 2 vols., London.

al-Biruni (1963). *Mineralogiya*, A. M. Byelenskiy, trans., Moscow.

Boltwood, Bertram (1907). On the Ultimate Disintegration Products of the Radio-Active Elements. Part II. The Disintegration Products of Uranium, *Amer. Jour. Sci.*, Vol. 23, pp. 77–88.

Borelli, Giovanni Alfonso (1670). Historia et meteorologia incendii Aetnaei anni 1669 . . . , Regio Iulio.

Born, Baron Inigo (1777). *Travels through the Bannat of Temeswar, Transylvania, and Hungary, in the Year 1770* . . . , R. E. Raspe, trans. & ann., Kearsley, London.

Bouguer, Pierre (1749). *La figure de la terre* . . . , Jombert, Paris.

Bournon, Jacques-Louis Comte de (1785). *Essai sur la lithographie de St. Étienne-en-Forez et sur l'origine de ses charbons de pierre*. Paris.

Bournon, Jacques-Louis Comte de (1808*a*). *Traité complet de la chaux carbonatée et de l'arragonite*, 3 vols., Phillips. London.

Bournon, Jacques-Louis Comte de (1808*b*). *Traité de minéralogie*, 3 vols., London.

Bournon, Jacques-Louis Comte de (1817). *Catalogue de la collection minéralogique particulière du roi*, 2 vols., Paris.

Boyle, Robert (1666). *The Origine of Forms & Qualities*, printed by H. Hall for R. Davis, Oxford.

Boyle, Robert (1672). *An Essay About the Origine & Virtues of Gems*, Pitt, London; reprinted Hafner, N.Y. (1972).

Boyle, Robert (1673). Reissue of Boyle (1672) with other essays.

Brewster, David (1814). On the Affections of Light Transmitted Through Crystallized Bodies, *Phil. Trans. of the Royal Soc.*, Vol. 104, pp. 187–218.

Briggs, Henry (1624). *Arithmetica logarithmica*, Jones, London.

Buch, Leopold von (1802–1809). *Geognostiche Beobachtungen auf Reisen durch Deutschland und Italien*, 2 vols., Hande & Spener, Berlin.

Buch, Leopold von (1809). On Volcanos and Craters of Elevation, *Edinburgh New Phil. Jour.*, Vol. XXI, p. 206.

Buch, Leopold von (1867–1885). *Leopold von Buch's gesammelte Schriften*, 4 vols., J. Ewald, J. Roth and H. Eck, eds. Berlin.

Buckland, William (1820). *Vindiciae geologicae; or the Connexion of Geology with Religion Explained* . . . . University Press, Oxford.

Buckland, William (1823). *Reliquiae Diluvianae; or Observations on the Organic Remains* . . . , Murray, London.

Buckland, William (1836). *Geology and Mineralogy Considered with Reference to Natural Theology*, 2 vols., Pickering, London.

Buffon, [Georges-Louis Leclerc], Comte de (1749–). *Histoire naturelle, générale et particulière*, Imprimerie Royale, Paris.

Buffon, [Georges-Louis Leclerc], Comte de (1778). *Époques de la nature*, Imprimerie Royale, Paris.

Buffon, [Georges-Louis Leclerc], Comte de (1792). *Barr's Buffon. Buffon's Natural History . . .*, Barr, London.

Bullard, Edward (1976). Epic Struggle with Geological Establishment, *Nature*, Vol. 259, p. 161.

Burnet, Thomas (1681). *Telluris theoria sacra . . .*, London.

Burnet, Thomas (1684). *The Sacred Theory of the Earth . . . The Two First Books . . .*, Kettilby, London.

Burnet, Thomas (1689). *Telluris theoria sacra . . . libri duo posteriores*, Kettilby, London.

Burnet, Thomas (1690). *An Answer to the Late Exceptions Made by Mr. Erasmus Warren Against the Theory of the Earth*, Kettilby, London, 86 p.

Burnet, Thomas (1691). *A Short Consideration of Mr. Erasmus Warren's Defence of his Exceptions against the Theory of the Earth, in a Letter to a Friend*, Kettilby, London, 42 p.

Cailleux, Andre (1979). The Geological Map of North America (1752) of J. E. Guettard, in C. Schneer, ed., *Two Hundred Years of American Geology*, Univ. Press of New England, Hanover.

Cajori, Florian (1929). *The Chequered Career of Ferdinand Rudolph Hassler, First Superintendent of the United States Coast Survey . . .*, Christopher, Boston.

Caley, Earle A. and John P. C. Richards (1956). *Theophrastus on Stones*, trans. & eds. Ohio State Univ. Press, Columbus.

Cardano, Girolamo (1550). *De subtilitate liber XXI*, Nuremberg.

Carozzi, Albert V. (1964). See Lamarck, 1802*a*.

Carozzi, Albert V. (1968). *Telliamed or Conversations Between an Indian Philosopher and a French Missionary on the Diminution of the Sea*, trans. & ed., Univ. of Illinois Press, Urbana.

Carswell, John (1950). *The Prospector . . .*, Cresset, London; published in the United States as *The Romantic Rogue*.

Cassini, Jacques (1722). *De la grandeur et de la figure de la terre*, Suite des Mémoires de l'Académie royale des sciences, Année 1718, Paris.

Cavendish, Henry (1798). Experiments to Determine the Density of the Earth, *Phil. Trans. of the Royal Soc. of London*, Vol. 88, pp. 469–526.

Ceruti, Benedetto and Andrea Chiocco (1622). Mvsaevm Franc. Calceolari[i] Ivn. Veronensis, [Verona].

Chenevix, Richard (1808). Sur quelques méthodes minéralogiques, *Annales de chimie*, Vol. 65, pp. 5–43, 113–160, 225–277.

Chenevix, Richard (1811). *Observations on Mineralogical Systems*, L. Horner, trans. Johnson, London.

Clairaut, Alexis-Claude (1743). *Théorie de la figure de la terre, tirée des principes de l'hydrostatique*, Paris.

Clark, John Willis, and Thomas McKenny Hughes (1890). *Life and Letters of the Reverend Adam Sedgwick*, 2 vols., Cambridge Univ. Press.

Clarke, Edward Daniel (1810–1823). *Travels in Various Countries of Europe, Asia, and Africa*, 6 vols., Cadell & Davies, London.

Clarke, John Mason (1923). *James Hall of Albany, 1811–1898, Geologist and Paleontologist*, Albany; reprinted Arno, N.Y. (1978).

Cleaveland, Parker (1816). *An Elementary Treatise on Mineralogy and Geology . . .*, Cummings & Hilliard, Boston.

Cleaveland, Parker (1822). *An Elementary Treatise on Mineralogy and Geology . . . ,* 2 vols., 2d ed. Cummings & Hilliard, Boston.

Columna, Fabius (Fabio Colonna) (1616). *De Glossopetris Dissertatio, in Purpura,* Rome.

Conybeare, William Daniel (1822). Additional Notices on the Fossil Genera *Ichthyosaurus* and *Plesiosaurus, Trans. of the Geol. Soc. of London,* 2d series, Vol. 1, pt. 1, pp. 103–123.

Conybeare, William Daniel and H. T. De la Beche (1821). Notice of a Discovery of a New Fossil Animal, Forming a Link Between the *Ichthyosaurus* and the Crocodile . . . , *Trans. of the Geol. Soc. of London,* Vol. 5, pp. 558–594.

Conybeare, William Daniel and William Phillips (1822). *Outlines of the Geology of England and Wales . . . ,* Phillips, London.

Cooper, Thomas (1560). *Chronicle,* London.

Cooper, Thomas (1813). Geology, *Emporium Arts Sci., n.s., Vol. 3, pp. 412–425.*

Cooper, Thomas (1817). Advertisement for Mineralogy Lectures, Univ. of Pennsylvania, *Analectic Magazine,* Vol. 10, p. 352.

Cope, Edward Drinker (1875). The Vertebrata of the Cretaceous Formations of the West, *Rpt. of the U.S. Geol. and Geog. Survey of the Territories (Hayden Survey),* Vol. II. U.S. Government Printing Office, Washington.

Cope, Edward Drinker (1884). The Vertebrata of the Tertiary Formations of the West, *Rpt. of the U.S. Geol. and Geog. Survey of the Territories (Hayden Survey),* Vol. III. U.S. Government Printing Office, Washington.

Cope, Edward Drinker (1886). *The Origin of the Fittest—Essays on Evolution,* Appleton, N.Y.

Copernicus, Nicolaus (1543). *De revolutionibus orbium coelestium,* libri VI. Petreius, Nuremberg.

Craig, G. Y., D. B. McIntire, and C. D. Waterston (1978). *James Hutton's Theory of the Earth: The Lost Drawings,* Scottish Academic Press, Edinburgh.

Croft, Herbert (1685). *Some Animadversions Upon a Book Intituled The Theory of the Earth,* Harper, London.

Cuvier, Georges (1811). *Essai sur la géographie minéralogique des environs de Paris, avec une carte géognostique, et des coupes de terrain,* Baudouin, Paris.

Cuvier, Georges (1812*a*). *Recherches sur les ossemens fossiles . . . ,* 4 vols., Paris.

Cuvier, Georges (1812*b*). Discours sur les révolutions du globe (Discours préliminaire of above).

Cuvier, Georges (1813). Essay on the Theory of the Earth . . . , with notes by Robert Jameson, Robert Kerry, trans. & ed., Blackwood, Edinburgh.

Cuvier, Georges (1818). *Essay on the Theory of the Earth . . . ,* Robert Jameson, trans. & ed., Samuel Latham Mitchill, ann., Kirk & Mercein, N.Y.

Cuvier, Georges (1825). *Recherches sur ossemens fossiles . . . ,* Troisième Édition, Chez G. Dufour et E. D'Ocagne, Paris.

Cuvier, Georges and A. Brongniart (1808–1811). Essai sur la géographie minéralogique des environs de Paris, *Phil. Mag.,* Vol. 35.

Dana, James Dwight (1837). *System of Mineralogy . . . ,* Herrick & Peck, New Haven.

Dana, James Dwight (1846). *Zoophytes, U.S. Exploring Expedition . . .* (Wilkes Expedition), with atlas, Vol., VII, Sherman, Philadelphia.

Dana, James Dwight (1849*a*). *Geology, U.S. Exploring Expedition . . .* (Wilkes Expedition), Vol. X, Sherman, Philadelphia.

Dana, James Dwight (1849*b*). *Zoophytes, atlas, U.S. Exploring Expedition . . .* (Wilkes Expedition), Sherman, Philadelphia.

Dana, James Dwight (1852–1853). *Crustacea, U.S. Exploring Expedition . . .* (Wilkes Expedition), Vol. XIII, XIV, Sherman, Philadelphia.

Dana, James Dwight (1855). *Crustacea, atlas, U.S. Exploring Expedition . . .* (Wilkes Expedition), Sherman, Philadelphia.

Darnton, Robert (1979). *The Business of Enlightenment, A Publishing History of the Encyclopédie, 1775–1800,* Harvard University Press, Cambridge, (Mass.).

Darwin, Charles (1842). *The Structure and Distribution of Coral Reefs,* Scott, London.

Darwin, Charles (1859). *On the Origin of Species . . . ,* Murray, London.

Daubeny, Charles (1826). *A Description of Active and Extinct Volcanos . . . ,* Phillips, London; Parker, Oxford.

Daubeny, Charles (1839). *Sketch of the Geology of North America . . . ,* Read before Ashmolean Soc., Nov. 26, 1836, Oxford.

Davies, Gordon L. (1969). *The Earth in Decay, a History of British Geomorphology 1578–1878,* Elsevier, N.Y.

Da Vinci, Leonardo (1938). *The Notebooks . . . ,* 2 vols., Edward MacCurdy, trans. & arr. Reynal & Hitchcock, N.Y.

Davy, Humphry (1830). *Consolations in Travel . . . ,* Murray, London.

Davy, Humphry (1980). *Humphry Davy on Geology . . . ,* Robert Siegfried and Robert H. Dott, Jr., ed. & ann., Univ. of Wisconsin Press, Madison.

Dawson, John William (1859). On Fossil Plants from the Devonian Rocks of Canada, *Geol. Soc. of London Quarterly Journal,* Vol. 15, pp. 477–488.

Dawson, John William (1888). *The Geological History of Plants,* Internat. Sci. Series, Vol. 61, Appleton, N.Y.

De la Beche, Henry (1839). *Report on the Geology of Cornwall, Devon and West Somerset,* Longman et al., London.

DeLuc, Jean André (1809). *Elementary Treatise on Geology . . . ,* Henry de la Fite, trans. Rivington, London.

Derham, W. (1718). *Philosophical Letters between the Late Learned Mr. Ray and Several of his Ingenious Correspondents, Natives and Foreigners . . . ,* Innys, London.

Descartes, René (1637). *Discours de la methode,* Leyden.

Descartes, René (1644). *Principia philosophiae,* Amsterdam.

Descartes, René (1681). *Les principes de la philosophie,* 4th ed., Paris.

Descartes, René (1911). *The Philosophical Works of Descartes,* Elizabeth S. Haldane and G.R.T. Ross, trans., Cambridge Univ. Press; reprinted Dover, N.Y. (1931).

Descartes, René (1971). *Oeuvres de Descartes,* Charles Adam and Paul Tannery, eds. Vrin, Paris.

Deshayes, Gerard Paul (1824–1837). *Description des coquilles fossiles des environs de Paris,* 2 vols., Paris.

Desmarest, Nicholas (1774). *Mémoire sur l'origine et la nature du basalte . . . ,* Mém. de l'Académie royale des sciences for 1771, Paris, pp. 705–775.

Desmarest, Nicholas (1777). *Mémoire sur le basalte.* Troisième partie . . . , Mem. de l'Académie royale des sciences for 1773, pp. 599–670.

Desmarest, Nicholas (1804). *Sur la constitution physique des couches de la colline de Montmarte . . . ,* Memoires de L'Institut des Sciences, Lettres, Arts, Paris.

Desmarest, Nicholas (1806). *Mémoire sur la détermination de trois époques de la nature par les produits des volcans . . .*, Mémoires de l'Institut des Sciences, Lettres, Arts, Paris.

Dewey, Chester (1819). A Geological Map of the Northwest Part of Massachusetts, *Amer. Jour. Sci.*, Vol. 1, no. 4, pp. 337–346.

Dewey, Chester (1824). Additional Remarks on the Geology of a Part of Massachusetts . . . , *Amer. Jour. Sci.*, Vol. 8, no. 2, pp. 240–244.

Diderot, Denis and Jean le Rond d'Alembert (1751–1765). *Encyclopédie, ou Dictionnaire raisonné des sciences, des arts et des métiers*, Paris (Neuchâtel).

Diderot, Denis and Jean le Rond d'Alembert (1762–1772). *Recueil de mille planches . . .*, (illustrations for the Encyclopédie), 11 vols.

Dolomieu, Dieudonné de Gratet de (1783). *Voyage aux îles de Lipari, fait en 1781 . . .*, Hôtel Serpente, Paris.

Dolomieu, Dieudonné de Gratet de (1788). *Mémoire sur les îles ponces, et catalogue raisonné des produits de l'Etna; . . .*, Cuchet, Paris.

Dolomieu, Dieudonné de Gratet de (1791). *Mémoire sur les pierres composées et sur les roches, Observations sur la physique, sur l'histoire naturelle et sur les arts*, Vol. 39, pp. 374–407.

Duhem, Pierre (1906–1913). *Études sur Leonard de Vinci . . .*, 3 vols., Paris (repro. 1955).

Dutton, C. E. (1880). Report on the Geology of the High Plateaus of Utah, with atlas, *U.S. Geog. & Geol. Survey, Rocky Mtn. Region* (Powell Survey), Washington.

Dutton, C. E. (1882). The Tertiary History of the Grand Canyon District, *U.S. Geol. Survey Monograph II*, Government Printing Office, Washington.

Eaton, Amos (1824). *A Geological and Agricultural Survey of the District Adjoining the Erie Canal, in the State of New-York*, Packard & Van Benthuysen, Albany.

Edmonds, J. M. (1978). *The Fossil Collection of the Misses Philpot of Lyme Regis*, *Proc. Dorset Nat. Hist. & Archaeol. Soc.*, Vol. 98, pp. 43–48.

Ellenberger, François (1975–1977). A l'aube de la géologie moderne: Henri Gautier (1660–1737), *Histoire et Nature*, nos. 7 & 9–10, 58+149 p.

Ellenberger, François (1978). Précisions nouvelles sur la découverte des volcans de France: Guettard, ses prédécesseurs, ses émules clermontois, *Histoire et Nature*, no. 12/13, pp. 3–42.

Elster, J.P.L.J. and F.K.H. Geitel (1902). Über eine fernere Analogie in dem elektrischen Verhalten der natürlichen und der durch Becquerelstrahlen abnorm leitend gemachten Luft, *Physikal. Zeit.*, Vol. 2., pp. 590–593.

Emmons, Ebenezer (1842). Geology of the Second District, in Geology, pt. 4, *Natural History of New York*, Carroll & Cook, Albany.

Emmons, Ebenezer (1844). *The Taconic System . . .*, Carroll & Cook, Albany.

Emmons, Ebenezer (1846). Agriculture of New York, Natural History of New York, Van Benthuysen, Albany.

Emmons, S. F. (1886). Geology and Mining Industry of Leadville, Colorado, *U.S. Geol. Survey Monograph* 12 (atlas 1887), Government Printing Office, Washington.

Ercker, Lazarus (1574). *Beschreibung allerfürnemisten mineralischen Ertz und Berckwerks arten . . .*, Schwartz, Prague.

Ercker, Lazarus (1951). *Treatise on Ores and Assaying, from the German Edition of 1580*, Anneliese Grünhaldt Sisco and Cyril Stanley Smith, trans. & ann., Univ. of Chicago Press.

Escholt, Mickel Pederson (1657). *Geologia norvegica*, Christiania (Oslo).

'Espinasse, Margaret (1956). *Robert Hooke*, The Contemporary Science Series, Heinemann, Toronto.

Eyles, Joan M. (1969*a*). William Smith (1769–1839) . . . , *Jour. Soc. Bib. Nat. Hist.*, Vol. 5, London, pp. 87–109.

Eyles, Joan M. (1969*b*). William Smith (1769–1839). A Chronology of Significant Dates in his Life, *Proc. Geol. Soc. London*, no. 1657, pp. 173–176.

Eyles, Joan M. (1973). Robert Jameson, *Dictionary of Scientific Biography*, Vol. 7, Scribners, N.Y.

Eyles, Victor (1971). John Woodward . . . a Bio-Bibliographical Account . . . , *Jour. Soc. Bibl. Nat. Hist.*, Vol. 5, no. 6, London, pp. 399–425.

Eyles, Victor (1971). H. T. De la Beche, *Dictionary of Scientific Biography*, Vol. 4, Scribners, N.Y.

Faujas de Saint-Fond, Barthélemy (1778). *Recherches sur les volcans éteints du Vivarais et du Velay*, Cuchet, Grenoble.

Faujas de Saint-Fond, Barthélemy (1797). *Voyage en Angleterre, en Écosse et aux Îles Hébrides* . . . , 2 vols., Jansen, Paris.

Faul, Henry (1959). Doubts of the Paleozoic Time Scale, *Jour. of Geophysical Research*, Vol. 64, p. 1102.

Faul, Henry (1960). Geologic Time Scale, *Bull. Geol. Soc. Am.*, Vol. 71, pp. 637–644.

Faul, Henry (1978). A History of Geologic Time, *Amer. Scientist*, Vol. 66, pp. 159–165.

Ferber, John James (1776). *Travels through Italy, in the Years 1771 and 1772* . . . , R. E. Raspe, trans., Davis, London.

Fisher, Osmond (1881). *Physics of the Earth's Crust*, Macmillan, London.

Fisher, Osmond (1882). On the Physical Cause of the Ocean Basins, *Nature*, Vol. 25, pp. 243–244.

Fitton, William H. (1813). On the Geological System of Werner, *Nicholson's Jour. of Nat. Phil.*, Vol. 36.

Fitton, William H. (1839). A Review of Mr. Lyell's "Elements of Geology," with Observations on the Progress of the Huttonian Theory of the Earth, *Edinburgh Review*, Vol. 69, pp. 406–440.

Foster, J. W. and J. D. Whitney (1850). *Report on the Geology and Topography of a Portion of the Lake Superior Land District, in the State of Michigan*, 2 vols., House Documents, Washington.

Franklin, Benjamin (1905–1907). *The Writings of Benjamin Franklin*, A. H. Smyth, ed., 10 vols., N.Y.

Frémont, John Charles (1845). *Report of the Exploring Expedition to the Rocky Mountains in the Year 1842, and to Oregon and North California in the Years 1843–44*, Senate, Executive Document 174, 693 p., Washington.

Fregoso, Battista (1509). *De Dictis Factisque Memorabilibus Collectanea*, Ferrarius, Milan.

Fyodorov, Evgraf Stepanovich (1891). *Simmetriia pravilnykh sistem figor, Zap. Min. Obshch.* (The Symmetry of Real Systems of Configurations, Trans. of the Min. Soc.) Vol. 28, pp. 1–146.

Galilei, Galileo (1632). *Dialogo . . . sopra i due Massimi Sistemi del Mondo Tolemaico, e Copernicano, Landini*, Florence.

Gautier, Henri (1721). *Nouvelles conjectures sur le globe de la Terre*, Paris.

Geikie, Archibald (1882). *Geological Sketches at Home and Abroad*, Macmillan, New York.

Geikie, Archibald (1905). *The Founders of Geology*, 2nd ed., Macmillan, London; reprinted, Dover, N.Y. n.d.

Gellibrand, Henry (1635). *A Discovrse Mathematical on the Variation of the Maneticall Needle. Together with Its Admirable Diminution Lately Discovered*, Jones, London.

Gercke, A. (1907). *Quaestiones Naturales*, Leipzig.

Gesner, Konrad (1565). *De Rervm Fossilivm*, Tiguri (Zurich).

Gilbert, Grove Karl (1877). Report on the Geology of the Henry Mountains, *U.S. Geog. & Geol. Survey of the Rocky Mtn. Region (Powell Survey)*, Government Printing Office, Washington.

Gilbert, Grove Karl (1890). Lake Bonneville, *U.S. Geol. Survey*, Monograph 1, Government Printing Office, Washington.

Gilbert, William (1600). *De magnete, magneticisque corporibus, et de magno magnete tellure . . .*, Short, London.

Goetzmann, William H. (1966). *Exploration and Empire . . .*, Knopf, N.Y.

Gould, Stephen J. (1979). Agassiz' Later Private Thoughts on Evolution . . . , in C. Schneer, ed., *Two Hundred Years of Geology in America*, Univ. Press of New England, Hanover.

Goyon, Georges (1949). Le papyrus de Turin dit 'des mines d'or et le Wadi Hammamat, *d'annales Service d'antiquité d'Égypte* (Cairo), Vol. 49, pp. 337–392.

Gressly, Amanz (1838). Observations géologiques sur le Jura Soleurois, *Neue Denkschriften der Schweizerischen naturforschenden Gesellschaft*, Vol. 2, pp. 1–112.

Greenough, George Bellas (1819). *Critical Examination of the First Principles of Geology*, Longman et al., London.

Guettard, Jean-Étienne (1751). Mémoire et carte minéralogique sur la nature et la situation des terrains qui traversent la France et l'Angleterre, *Mém. Acad. Roy. Sci. Paris (1746)*, pp. 363–392.

Guettard, Jean-Étienne (1752*a*). Carte minéralogique où l'on voit la nature des terrains du Canada et de la Louisiane, Plate 7 in *Mémoire dans lequel on compare le Canada à la Suisse par rapport à ses minéraux, Hist. Acad. Roy. Sci. Paris*, Vol. 4, p. 189.

Guettard, Jean-Étienne (1752*b*). Mémoire dans lequel on compare le Canada à la Suisse, par rapport à ses minéraux, *Hist. Acad. Roy. Sci. Paris.*

Guettard, Jean Étienne (1755). Mémoire sur les Granits de France, Comparés à ceux d'Égypte, *Mém. Acad. Roy. Sci. Paris* (1751), pp. 164–210.

Guettard, Jean-Étienne (1756). Mémoire Dans Lequel on Compare le Canada à la Suisse, par Rapport à ses Minéraux, *Mém. Acad. Roy. Sci. Paris* (1752), pp. 189–220.

Guettard, Jean-Étienne (1764). Mémoire sur la Nature du Terrain de la Pologne et des Minéraux qu'il Renferme, *Mém. Acad. Roy. Sci. Paris*, (1762), pp. 234–257, 293–336.

Guettard, Jean-Étienne (1779). *Memoires sur la minéralogie au Dauphine*, 2 vols., Paris.

Gunther, R. T. (1937–). *Early Science in Oxford*, 14 vols., University Press, Oxford.

Gutenberg, Beno (1913). *Die seismische Bodenunruhe*, Beitr-Geophysik Leipzig, Vol. 11, pp. 314–353.

Gutenberg, Beno (1927). Die Veränderungen der Erdkruste durch Fliessbewegungen der Kontinentalscholle, *Gerlands Beiträge zur Geophysik*, Vol. 16, pp. 239–247, Vol. 18, pp. 281–291.

Hague, James Duncan (1870). Mining Industry, *U.S. Geol. Expl. of the 40th Parallel (King Survey)*, Vol. III, Government Printing Office, Washington.

Hague, James Duncan and S. F. Emmons (1877). Descriptive Geology, *U.S. Geol. Expl. of the 40th Parallel (King Survey)*, Vol. II, Government Printing Office, Washington.

Hailstone, John (1792). *A Plan of a Course of Lectures on Mineralogy*, Cambridge.

Haldane, Elizabeth S. and G.R.T. Ross (1911). *The Philosophical Works of Descartes*, 2 vols., Cambridge University Press.

Hall, Courtney Robert (1934). *A Scientist in the Early Republic, Samuel Latham Mitchill, 1764–1831*, Columbia University Press, N.Y.

Hall, Sir James (1800). Experiments upon Whinstone and Lava, *Nicholson's Jour. Nat. Phil. Chem. Arts*, Vol. 4, pp. 8–18, 56–65.

Hall, Sir James (1806). Account of a Series of Experiments Shewing the Effects of Compression in Modifying the Action of Heat, *Nicholson's Jour. Nat. Phil. Chem. Arts*, Vol. 13, pp. 328–343, Vol. 14, pp. 13–22, 113–128, 196–212, 302–318.

Hall, James (1843). *Geology of New York*, pt. IV; Comprising the Survey of the Fourth Geological District, Carroll & Cook, Albany.

Hallam, A. (1973). *A Revolution in the Earth Sciences*, Clarendon Press, Oxford.

Haller, Albrecht von (1756). *Physikalische Betrachtungen von den Erdbeben . . .* , Brönner, Frankfurt & Leipzig.

Hamilton, William (1776). *Campi phlegraei . . .* , 2 vols., Naples, 90 p.

Hamilton, William (1779). *Supplément au Campi phlegraei*, Naples.

Harrington, John W. (1967). The First, First Principles of Geology, *Amer. Jour. Sci.*, Vol. 265, pp. 449–461.

Hartmann, Philip Jacob (1677). *Succini Prussici Physica & Civilis Historia*, Hallervord, Frankfurt.

Haüy, René-Just (1784). *Essai d'une théorie sur la structure des crystaux . . .* , Chez Gogué & Née de la Rochelle, Paris.

Hawkins, Thomas (1840). *The Book of the Great Sea-Dragons, Ichthyosauri and Plesiosauri . . .* , Pickering, London.

Hayden, Ferdinand Vandeveer (1877). Geol. & Geog. Atlas of Colorado . . . , *U.S. Geol. & Geog. Survey of the Territories (Hayden Survey)*, Government Printing Office, Washington.

Hayden, Horace H. (1820). *Geological Essays . . .* , Robinson, Baltimore.

Hazen, Robert M. and Margaret Hindle Hazen (1980). *American Geological Literature, 1669 to 1850*, Dowden, Hutchinson & Ross, Stroudsburg (Pa.).

Heer, Oswald (1876). *The Primaeval World of Switzerland*, 2 vols., Longmans, Green, London.

Hennepin, Louis (1697). *Nouvelle Découverte d'un tres Grand Pays Situé dans l'Amérique . . .* , Utrecht.

Herodotus (1858–1860). *The History of Herodotus*, George Rawlinson, trans. & ed., 4 vols., London.

Hitchcock, Edward (1833). *Report on the Geology, Mineralogy, Botany, and Zoology of Massachusetts . . .* , with an atlas. Adams, Amherst.

d'Holbach, Paul Henri Thiry, Baron (1770). *Système de la nature*, London (Amsterdam).

Hölder, Helmut (1960). *Geologie and Paläontologie in Texten und ihrer Geschichte*, Alber, Freiburg/Munich.

Holmes, Arthur (1911). The Association of Lead with Uranium in Rock Minerals, and its Application to the Measurement of Geological Time, *Proc. Roy. Soc.*, series A, Vol. 85, pp. 248–256.

Holmes, Arthur (1913). *The Age of the Earth*, Harper, N.Y.

Holmes, Arthur (1927). *The Age of the Earth, an Introduction to Geological Ideas*, Benn, London.

Holmes, Arthur (1928). Continental Drift (review of the symposium volume), *Nature*, Vol. 122.

Holmes, Arthur (1928–1929). Radioactivity and Earth Movements, *Trans. Geol. Soc. of Glasgow*, Vol. 18, pt. 3, pp. 559–606.

Holmes, Arthur (1937). *The Age of the Earth*, new ed. Nelson, London.

Holmes, Arthur (1944). *Principles of Physical Geology*, Ronald Press, N.Y.

Holmes, Arthur (1947). The Construction of a Geological Time-Scale, *Trans. Geol. Soc. Glasgow*, Vol. 21, pp. 117–152.

Holmes, Arthur (1960). A Revised Geological Time-Scale, *Edinburgh Geol. Soc. Trans.*, Vol. 17, pt. 3, pp. 183–216.

Holmes, Arthur (1965). *Principles of Physical Geology* (revised), Ronald Press, New York.

Holmyard, E. J. and D. C. Mandeville (1927). *The Book of the Remedy (Avicenna)*, Geuthner, Paris.

Holland, Philemon, trans. (1601). *Pliny's Natural History*, reprinted 1634, London.

Hooke, Robert (1665). *Micrographia . . .* , Martyn & Allestry, London; reprinted Dover, N.Y. (1961).

Hooke, Robert (1705). *The Posthumous Works of Robert Hooke*, Richard Waller, ed. Smith & Walford, London.

Hugenius, Christian (1673). *Horologivm oscillatorivm sive de motu pendulorvm ad horologia aptato demonstrationes geometricae*, Muguet, Paris.

Hume, W. F. (1937). *Geology of Egypt*, Vol. 2, The Fundamental Pre-Cambrian Rocks of Egypt and the Sudan; their Distribution, Age, and Character, Government Press, Cairo.

Hunt, Thomas Sterry (1872). History of the Names Cambrian and Silurian in Geology, *Can. Nat., n.s. 6., pp. 281–312, 417–448.*

Hutton, Charles (1779). An Account of the Calculations made from the Survey and Measures taken at Schehallien . . . in Order to Ascertain the Mean Density of the Earth, *Phil. Trans. Royal Soc.*, Vol. 68, pp. 689–778.

Hutton, James (1785). *Abstract of a Dissertation Read in the Royal Society of Edinburgh. . .* , np, nd.

Hutton, James (1788). Theory of the Earth; or an Investigation of the Laws Observable in the Composition, Dissolution, and Restoration of Land upon the Globe, *Trans. Royal Soc. Edinburgh.*, Vol. 1, pp. 209–304.

Hutton, James (1899). *Theory of the Earth, with Proofs and Illustrations*, A. Geike, ed., Vol. 3, Burlington House, London.

Huxley, Thomas (1869). Anniversary Address of the President, *Geol. Soc. London Quart. Jour.*, pp. 38–53.

Huygens, Christian (1673). See Hugenius.

Hyde, T., Ed. (1665). *Tabulae longitudinis et latitudinis stellarum fixarum ex observatione Ulagh Beighi (Azophi)*, Oxford.

Ibn Sina (1927). *See* Holmyard and Mandeville.

Imperato, Ferrante (1599). *Dell' Historia Natvrale* . . . Libri XXVIII, Vitale, Naples.

Irving, Washington (1837). *The Adventures of Captain Bonneville, U.S.A.*, Philadelphia; 3 vols., Bentley, London.

Jahn, Melvin E. and Daniel Woolf (1963). *The Lying Stones of Dr. Bartholomew Adam Beringer* . . . , Univ. of California Press, Berkeley.

James, Thomas C. (1826). A Brief Account of the Discovery of Anthracite Coal on the Lehigh, *Memoirs of the Historical Soc. of Pennsylvania*, Vol. 1, pp. 315–320.

Jameson, Robert (1805). *Treatise on the External Characters of Minerals*, Univ. Press, Edinburgh.

Jameson, Robert (1808). *Elements of Geognosy*, being vol. III and part II of the System of Mineralogy, Edinburgh.

Jeffreys, Harold (1924). *The Earth, Its Origin, History, and Physical Constitution*, Cambridge Univ. Press.

Jeffreys, Harold (1929). *The Future of the Earth*, Norton & Co., N.Y.

Jovanović, Borislav (1971). Early C er Metallurgy of the Central Balkans, *Actes 8e Congres Internat. Sci. Pre'hist. Protohist*, Vol. 1, pp. 131–139.

Jukes, Joseph B. (1862). On the Mode of Formation of the River-Valleys in the South of Ireland, *Quart. Jour. Geol. Soc. London*, Vol. 18, pp. 378–403.

Jussieu, Antonine de (1718). Examen des causes des impressions de plantes marqueés sur certaines pierres des environs de St. -Chaumont dans le Lyonnais, *Mémoires de l'Académie royale des sciences*, pp. 287–297.

Jussieu, Antoinede de (1719). "Appendix" in J. P. de Tournefort *Institutions rei herbariae*, 3d ed., Paris.

Keill, John (1698). *An Examination of Dr. Burnet's Theory of the Earth. Together with Some Remarks on Mr. Whiston's New Theory of the Earth*, Theater, Oxford.

Keill, John (1699). *An Examination of the Reflections on the Theory of the Earth. Together with a Defence of the Remarks on Mr. Whiston's New Theory*, Clemens, Oxford.

Kelvin, Lord. See Thomson, William.

Kentmannus, Io. (1565). *Nomenclaturae Rerum fossilium, quae in Misnia praecipue & in aliis quoque regionibus inveniuntur*, Tiguri (Zurich).

Kettner, Radim (1963). Jean Louis Giraud Soulavie (1752–1813), *Cas. Miner. Geol.*, Vol. 8, n. 4, pp. 402–403.

Kettner, Radim (1967). *Geologické vědy na vysokých školách pražských*, Universita Karlova, Praha.

Kidd, John (1809). *Outlines of Mineralogy*, 2 vols., Parker, Oxford.

Kidd, John (1815). *A Geological Essay on the Imperfect Evidence in Support of a Theory of the Earth*, Oxford.

King, Clarence (1876). *Report Upon the Geological Exploration of the Fortieth Parallel . . . with Atlas*, Government Printing Office, Washington.

King, Clarence (1878). Systematic Geology, *U.S. Geol. Explor. of the 40th Parallel (King Survey)*, Vol. 1, Government Printing Office, Washington.

King, Clarence (1893). The Age of the Earth, *Amer. Jour. Sci.*, Vol. 45, pp. 1–20.

Kircher, Athanasius (1665). *Mundus subterraneus*, Jansson & Waesbergh, Amsterdam.

Kirk, G. S. and J. E. Raven (1962). *The Presocratic Philosophers*, University Press, Cambridge.

Kirwan, Richard (1793). Examination of the Supposed Igneous Origin of Stony Substances, *Trans. Royal Irish Acad.*, Vol. 5, pp. 51–81.

Kirwan, Richard (1794). *Elements of Mineralogy*, 2d ed., Vol. 1, London.

Kirwan, Richard (1802). Observations on the Proofs of the Huttonian Theory of the Earth Adduced by Sir James Hall, Bart., *Trans. Royal Irish Acad.*, Vol. 8, pp. 3–27.

Krejčí, Jan (1884). Joachim Barrande, *Čas. Mus. Král. Českého*, Vol. 58, pp. 385–404.

Lake, Philip (1922). Wegener's Displacement Theory, *Geological Mag.*, Vol. 59, pp. 338–346.

Lake, Philip (1923). Wegener's Hypothesis of Continental Drift, *Geographical Jour.*, Vol. 61, pp. 179–194.

Lamarck, Jean Baptiste Pierre Antoine de Monet (1802*a*). *Hydrogéologie* . . . , Paris; Albert Carozzi, trans. & ann. Univ. of Illinois Press, Urbana (1964).

Lamarck, Jean Baptiste Pierre Antoine de Monet (1802*b*). *Recherches sur l'organisation des corps vivants précédé du "Discours d'ouverture"* . . . , Paris.

Lamarck, Jean Baptiste Pierre Antoine de Monet (1815–1822). Histoire naturelle des animaux sans vertèbres, 7 vols., Verdiere, Paris.

Lamarck, Jean Baptiste Pierre Antoine de Monet (1964). *Hydrogeology*, Albert Carozzi, trans. & ann., Univ. of Illinois Press, Urbana.

Lapworth, Charles (1879). On the Tripartite Classification of the Lower Paleozoic Rocks, *Geol. Mag.*, n.s., Vol. 6, pp. 1–15.

La Rocque, Aurèle (1969). Bernard Palissy, in C. Schneer, ed., *Toward a History of Geology*, MIT Press, Cambridge (Mass.).

Leidy, Joseph (1873). Contributions to the Extinct Vertebrate Fauna of the Western Territories, *U.S. Geol. and Geog. Survey of the Territories (Hayden Survey)*, rpt. 1, Government Printing Office, Washington.

Leidy, Joseph (1879). Fresh-Water Rhizopods of North America, U.S. Geol. and Geog. Survey of the Territories (Hayden Survey), rpt. 12, Government Printing Office, Washington.

Lesley, J. Peter (1876). Historical Sketch of Geological Explorations in Pennsylvania and Other States, *2nd Geol. Survey of Pennsylvania*, Harrisburg.

Lesquereux, Leo (1858). *The Fossil Plants of the Coal Measures of the U.S. with Descriptions of the New Species in the Cabinet of the Pottsville Scientific Association*, Banna, Pottsville (Pa.).

Lesquereux, Leo (1874). Contributions to the Fossil Flora of the Western Territories, Part I, The Cretaceous Flora, rpt. 6, U.S. Geol. and Geog. Survey of the Territories (Hayden Survey), Government Printing Office, Washington.

Lesquereux, Leo (1878). Contributions to the Fossil Flora of the Western Territories, Part II, The Tertiary Flora, rpt. 7, *U.S. Geol. and Geog. Survey of the Territories (Hayden Survey)*, Government Printing Office, Washington.

Lesquereux, Leo (1883). Contributions to the Fossil Flora of the Western Territories, Part III, The Cretaceous and Tertiary Floras, rpt. 8, *U.S. Geol. and Geog. Survey of the Territories (Hayden Survey)*, Government Printing Office, Washington.

Lhwyd, Edward (1698). Several Regularly Figured Stones, *Phil. Trans. Royal Soc.*, Vol. 20, no. 243, pp. 279–280.

Lhwyd, Edward (1699). *Lithophylacii Britannici ichnographia*, London.

Lhwyd, Edward (1945). Early Science in Oxford, Vol. 14, *Life and Letters of Edward Lhwyd*, R. T. Gunther, ed., Oxford.

Lindgren, Waldemar (1913). *Mineral Deposits*, McGraw-Hill, New York.

Lister, Martin (1678). *Historiae animalium Angliae . . .*, Martyn, London.

Lister, Martin (1684). *De thermis et fontibus medicatis Angliae, Dettilly, London.*

Logan, William Edmond (1842). On the Character of the Beds of Clay Lying Immediately Below the Coal Seams of South Wales . . . , *Geol. Soc. of London, Trans.*, 2d ser., Vol. 6, p. 491.

Logan, William Edmond, Alexander Murray, T. Sterry Hunt, and E. Billings (1863). Geology of Canada, *Canadian Geological Survey*, 983 p.

Lough, John (1968). *Essays on the Encyclopédie of Diderot and d'Alembert*, Oxford University Press.

Lovell, Robert (1661). *An Universal History of Minerals, Containing the Summe of All Authors. . .*, Oxford.

Lucretius (1951). *Lucretius on the Nature of the Universe*, Penguin, Baltimore.

Lyell, Charles (1827*a*). State of the Universities, *Quart. Review*, Vol. 36, pp. 216–268.

Lyell, Charles (1827*b*). Memoir on the Geology of Central France. . . , *Quart. Review*, Vol. 36, pp. 437–483 (unsigned review of Scrope, 1827).

Lyell, Charles (1830). *Principles of Geology*, Murray, London.

Lyell, Charles (1833). *Principles of Geology*, 3 vols., Murray, London.

Lyell, Charles (1837). *Principles of Geology*, 2 vols., Kay, Philadelphia.

Lyell, Charles (1838). *Elements of Geology*, Murray, London.

Lyell, Charles (1841). *Elements of Geology*, reprinted from 2d English ed., 2 vols., Boston.

Lyell, Charles (1842). *Principles of Geology*, reprinted from the 6th English ed., Hilliard, Gray, & Co., Boston.

Lyell, Charles (1845). *Travels in North America, in the Years 1841–2 . . .*, Wiley & Putnam, N.Y.

Lyell, Charles and Roderick Impey Murchison (1829). On the Excavation of Valleys, as Illustrated by the Volcanic Rocks of Central France, *Edin. New Phil. Jour.*, Vol. 7, pp. 15–48.

McAllister, Ethel (1941). *Amos Eaton, Scientist and Educator, 1776–1842*, Univ. of Pennsylvania, Philadelphia.

Macculloch, John (1814). On the Granite Tors of Cornwall, *Trans. Geol. Soc. London*, Vol. 2, pp. 66–78.

Macculloch, John (1819). *A Description of the Western Islands of Scotland . . .*, 3 vols., Constable, Edinburgh; Hurst, Robinson & Co., London.

Macculloch, John (1821). *A Geological Classification of Rocks . . .*, London.

Macculloch, John (1831). *A System of Geology, with a Theory of the Earth*, 2 vols., London.

Macculloch, John (1836). *A Geological Map of Scotland . . .*, Arrowsmith, London.

Maclure, William (1809*a*). Observations on the Geology of the United States, Explanatory of a Geological Map, *Trans. Amer. Phil. Soc.*, Vol. 6, pt. 2, pp. 411–428.

Maclure, William (1809*b*). Observations on the Geology of the United States, Explanatory of a Geological Map, with colored geological map by Samuel B. Lewis, *Trans. Amer. Phil. Soc.*, Vol. 6, pt. 2, pp. 411–428.

Maclure, William (1817). *Observations on the Geology of the United States of America . . .* Accompanying Geological Map, printed for the author by A. Small, Philadelphia.

Maclure, William (1818*a*). Observations on the Geology of the United States of North America . . . , *Trans. Amer. Phil. Soc.*, Vol. 1, n.s., pp. 1–91.

Maclure, William (1818*b*). Essay on the Formation of Rocks, or an Inquiry into the Probable Origin of their Present Form and Structure, *Phila. Acad. Nat. Sci. Jour.*, Vol. 1, pt. 2, pp. 261–271.

Maillet, Benoît de (1735). *Description de l'Égypte . . .* , Genneau, Rollin, Paris.

Maillet, Benoît de (1748). *Telliamed ou entretiens d'un philosophe indian avec un missionnaire françois . . .* , Amsterdam.

Maillet, Benoît de (1968). *Telliamed or Conversations Between an Indian Philosopher and a French Missionary on the Diminution of the Sea*, Albert V. Carozzi, trans. & ed., Univ. of Illinois Press, Urbana.

Mallet, Robert (1849, 1851, 1859, 1871). *Admiralty Manual of Scientific Inquiry.*

Mallet, Robert (1850, 1851, 1852–1854, 1858). *Report(s) to the British Association.*

Mantell, Gideon (1822). *The Fossils of the South Downs, or Illustrations of the Geology of Sussex*, Relfe, London.

Mantell, Gideon (1825). On the Teeth of the Iguanodon . . . , *Phil. Trans. of the Royal Soc.*, Vol. 115, pp. 179–186.

Mantell, Gideon (1940). *The Journal of Gideon Mantell, Surgeon and Geologist*, E. C. Curwen, ed., Oxford.

Marsh, Othniel Charles (1880). Odontornithes, *U.S. Geol. Explor. of the 40th Parallel (King Survey)*, Vol. VII, Government Printing Office, Washington.

Marvin, Ursula B. (1973). *Continental Drift, the Evolution of a Concept*, Smithsonian Institution, Washington.

Maskelyne, Nevil (1775–1779). Accounts of Experiments and Calculations to Determine the Attraction of Schehallion . . . , extracted from v. 65, 68, 70, *Phil. Trans. of the Royal Soc.*

Mathesius, Johann (1562). *Sarepta oder Bergpostill Sampt der Jotchimsthalischen kurzen Chroniken*, Nürnberg.

Maupertuis, Pierre Louis Moreau de (1738). *La figure de la terre . . .* , Imprimerie Royale, Paris.

Maury, Matthew Fontaine (1855). *Physical Geography of the Sea*, Harper, N.Y.

Meek, F. B., James Hall, James Whitfield, and R. P. Ridgway (1877). Palaeontology, *U.S. Geol. Explor. of the 40th Parallel (King Survey)*, pt. 1, Vol. IV, pp. 1–197.

Mercati, Michele (1719). *Metallotheca Vaticana*, Rome.

Merrill, George P. (1924). *The First One Hundred Years of American Geology*, Yale University, New Haven; reprinted Hafner, N.Y. (1969).

Michell, John (1760). Conjectures Concerning the Cause and Observations upon the Phaenomena of Earthquakes, Vol. 51, pt. 2, *Phil. Trans. of the Royal Soc.*, pp. 566–634.

Michell, John (1818). Conjectures Concerning the Cause and Observations upon the Phaenomena of Earthquakes, reprint, *Phil. Mag.*

Milne, John (1886). *Earthquakes and Other Earth Movements*, Appleton, N.Y.

Mitchill, Samuel Latham (1792). *Outline of the Doctrines in Natural History, Chemistry, and Economics*, Childs & Swaine, N.Y.

Mitchell, Samuel Latham (1798). A Sketch of the Mineralogical History of the State of New York, *Med. Repos.*, Vol. 1, pp. 293–314, 445–452.

Mitchell, Samuel Latham (1802). Additional Articles of my Report to the Agricultural Society on the Mineralogy of New York, *Med. Repos.*, Vol. 5, pp. 212–215.

Morley, Henry (1852). *Palissy the Potter. The Life of Bernard Palissy* . . . , 2 vols., Chapman & Hall, London.

Moro, Anton-Lazzaro (1740). *Dei crostacei e degli altri corpi marini che si trovano sui monti libri due*, Ceremia, Venice.

Moscardo, L. (1656). *Note overo Memoirie del Museo suo, Padoa.*

Murchison, Roderick Impey (1939). *The Silurian System*, Murray, London.

Murchison, Roderick Impey, Edouard de Verneuil and Alexander von Keyserling (1845). *The Geology of Russia in Europe and the Ural Mountains*, 2 vols., Murray, London.

Murray, John (1802). *A Comparative View of the Huttonian and Neptunian Systems of Geology* . . . , Ross & Blackwood, Edinburgh; Longman, Rees, London.

Napier, John (1614). *Mirifici logarithmorum canonis descriptio ejusque usus* . . . , Edinburgh.

Nelson, Clifford M., Mary C. Rabbitt, and Fritiof M. Fryxell (1981). Ferdinand Vandeveer Hayden—The U.S. Geological Survey Years, 1879–1886, *Amer. Phil. Soc. Proc.*, Vol. 125, no. 3, pp. 238–243.

Newton, Henry and W. P. Jenney (1880). Report on the Geology and Resources of the Black Hills of Dakota, *U.S. Geol. & Geog. Survey of the Rocky Mountain Region (Powell Survey)*, 566 p.

Newton, Isaac (1704). *Opticks: or a Treatise of the Reflexions, Refractions, Inflexions and Colours of Light*, Smith & Walford, London.

Newton, Isacc (1729). *The Mathematical Principles of Natural Philosophy*, Andrew Motte, trans., 2 vols., Benjamin Motte, London.

Nicol, William (1829). On a Method of so Far Increasing the Divergency of the Two Rays in Calcareous Spar that Only One Image may be Seen at a Time, *Edinburgh New Phil. Jour.*, Vol. 6, pp. 83–84.

Nicolson, Marjorie Hope (1959). *Mountain Gloom and Mountain Glory*, Norton, N.Y.

Nier, Alfred O. (1939). *The Isotopic Constitution of Radiogenic Leads and the Measurement of Geologic Time II, Phys. Review*, Vol. 55, pp. 153–163.

Norman, Robert (1581). *The Newe Attractive, Containing a Short Discourse of the Magnes or Lodestone*, Ballard, London.

Oakley, Kenneth (1965). Folklore of Fossils, *Antiquity*, Vol. 39, pp. 9–16, 117–125, reprinted in *N.Y. Paleont. Soc. Notes*, Vol. 5, pp. 1–20, 1974.

Oldenburg, Henry (1965–1972). *The Correspondence of Henry Oldenburg*, A. Rupert Hall and Marie Boas Hall, ed. & trans., Univ. of Wisconsin Press, Madison, Milwaukee; London.

Oldham, Richard Dixon (1906). The Constitution of the Interior of the Earth . . . , *Phil. Mag.*, Vol. 12, pp. 165–166.

Olivi, J. Bapt. (1584). De reconditis et praecipuis collectaneis in museo calceolario asservatis . . . , *Trans. Linn. Soc. London*, Vol. VII, p. 230.

Oppenheim, A. Leo, Robert H. Brill, Dan Barag, and Axel Von Saldern (1970). *Glass and Glassmaking in Ancient Mesopotamia*, Corning, N.Y.

Otter, William (1824). *The Life and Remains of Edward Daniel Clarke*, 2 vols., London.

Owen, David Dale (1852). *Report of a Geological Survey of Wisconsin, Iowa, and Minnesota* . . . , Lippincott, Grambo, & Co., Philadelphia.

Owen, Richard (1857). *Key to the Geology of the Globe; an Essay*, Lippincott, Philadelphia.

Palissy, Bernard (1563). *Recette véritable par laquelle tous les hommes de la France pourront apprendre à multiplier et augmenter leurs trésors*, Berton, La Rochelle.

Palissy, Bernard (1580). *Discours admirables de la nature des eaux et fontaines, tant naturelles qu'artificielles, des métaux, des sels et salines, des pierres, des terres, . . .*, Martin le Jeune, Paris.

Palissy, Bernard (1957). *The Admirable Discourses of Bernard Palissy*, A. La Roque, trans. and ed., Univ. of Illinois Press, Urbana.

Parkinson, James (1804–1811). *Organic Remains of a Former World . . .*, 3 vols., London.

Parkinson, James (1822). *Outlines of Oryctology: an Introduction to the Study of Fossil Organic Remains . . .*, London.

Pausanias (1971). *Guide to Greece*, P. Levi, trans., 2 vols., Penguin, Baltimore.

Pease, Arthur Stanley (1942). Fossil Fishes Again, *Isis*, Vol. 33, pp. 689–690.

Peithner, Johann (1780). *Versuch über die naturliche und politische Geschichte der böhmischen und mährischen* Bergwerke, Wien.

Pettus, Sir John (1683). *Fleta Minor. The Laws of Art and Nature in Knowing, Judging, Assaying, Fining, Refining and Inlarging the Bodies of Confin'd Metals*, London.

Phillips, William (1811). A Description of the Red Oxyd of Copper . . . , *Trans. of the Geol. Soc. of London*, Vol. 2, pp. 23–37.

Phillips, William (1814). A Description of the Oxyd of Tin . . . , *Trans. of the Geol. Soc. of London*, Vol. 2, pp. 336–376.

Phillips, William (1815). Outlines of Mineralogy and Geology, *Trans. of the Geol. Soc. of London*.

Phillips, William (1816). *Elementary Introduction to the Knowledge of Mineralogy*, Phillips, London.

Phillips, William (1818). *A Selection of Facts from the Best Authorities . . . , Outline of the Geology of England and Wales*, Phillips, London.

Phillips, William and William Conybeare (1822). *Outlines of the Geology of England and Wales . . .*, London.

Placet, Francois (1666). *La corruption du grande et petit Monde*, Alliot, Paris.

Playfair, John (1802). *Illustrations of the Huttonian Theory of the Earth*, Cadell & Davies, Edinburgh.

Playfair, John (1805). Biographical Account of the Late James Hutton, F.R.S., Edinburgh, *Trans. Royal Soc. of Edinburgh*, Vol. 5, pt. 3, pp. 39–99.

Plot, Robert (1677). *The Natural History of Oxfordshire*, Theater, Oxford.

Plot, Robert (1686). *The Natural History of Staffordshire*, Theater, Oxford.

Plot, Robert (1705). *The Natural History of Oxfordshire*, 2d ed.

Porter, Roy (1977). *The Making of Geology, Earth Science in Britain, 1660–1815*, Cambridge Univ. Press.

Pott, Johann Heinrich (1746). *Lithogeognosia: Chymische Untersuchungen*, Pottsdam.

Powell, John Wesley (1875). *Exploration of the Colorado River of the West and Its Tributaries Explored in 1869, 1870, 1871, & 1872*, House Ex. Doc., Washington.

Powell, John Wesley (1876). Report on the Geology of the Eastern Portion of the Uinta Mountains and a Region of Country Adjacent Thereto, *U.S. Geol. & Geog. Survey of the Territories (Powell Survey)*, Vol. VII, Government Printing Office, Washington.

Powell, John Wesley (1879). *Report on the Lands of the Arid Region of the United States with a More Detailed Account of the Lands of Utah*, U.S. 45th Cong. 2d Session, House Ex. Doc. 73, 2d ed., Washington.

Poynting, J. H. (1913). *The Earth, Its Shape, Size, Weight, and Spin*, Cambridge Univ. Press.

Pumpelly, Raphael (1879). The Relation of Secular Rock Disintegration to Loess, Glacial Drift, and Rock Basins, *Am. Jour. Sci.*, Vol. 17, no. 3, pp. 133–144.

Pumpelly, Raphael (1894). Geology of the Green Mountains in Massachusetts, *U.S. Geol. Survey*, monograph 23, pp. 5–34.

Rabbitt, Mary (1979). *Minerals, Lands, and Geology for the Common Defence and General Welfare, Vol. 1, Before 1879*, Government Printing Office, Washington.

Rabbitt, Mary (1980). *Minerals, Lands, and Geology for the Common Defence and General Welfare, V. 2., 1879–1904*, Government Printing Office, Washington.

Rappaport, Rhoda (1960). G. F. Rouelle: an Eighteenth Century Chemist and Teacher, *Chymia*, Vol. 6, pp. 68–101.

Rappaport (1969). The Geological Atlas of Guettard, Lavoisier, and Monnet . . . , in C. Schneer, ed., *Toward a History of Geology*, MIT Press, Cambridge (Mass.), pp. 272–287.

Raspe, Rudolf Erich (1763). *Specimen Historiae Naturalis Globi Terraquei, praecipue de novis a mari natis insulis . . .* , Schreuder & Mortier, Amsterdam and Leipzig.

Raspe, Rudolf Erich (1771). Nachricht von einigen nieder hessischen Basalten . . . , *Deutschen Schriften der Kgl.* Societät der Wissenschaften in Göttingen, Vol. 1.

Raspe, Rudolf Erich (1776). *An Account of Some German Volcanoes*, Davis, London.

Ray, John (1670 &c.). *Catalog of English Plants*, Cambridge.

Ray, John (1691). *The Wisdom of God Manifested in the Works of the Creation*, Smith, London.

Ray, John (1692). *Miscellaneous Discourses Concerning the Dissolution and Changes of the World . . .* , Smith, London.

Ray, John (1693). *Three Physico-Theological Discourses . . .* , 2d ed., Smith, London.

Ray, John (1713). *Three Physico-Theological Discourses . . .* , 3d ed., Innys, London.

Ray, John (1718). *Philosophical Letters Between the Late Learned Mr. Ray and Several of his Ingenious Correspondents, Natives and Foreigners . . .* , W. Derham, ed., Innys, London.

Ray, John (1721). *Three Physico-Theological Discourses . . .* , 4th ed., Innys, London.

Ray, John (1732). *Three Physico-Theological Discourses . . .* , 4th ed., corrected, Innys, London.

Ray, John (1846). *The Correspondence of John Ray . . .* , Edwin Lankester, ed., Ray Society, London.

Ray, John (1928). *Further Correspondence of John Ray*, Robert W. T. Gunther, ed., Ray Society, London.

Réaumur, René-Antoine Ferchault de (1722). *L'art de convertir le fer forgé en acier . . .* , Paris, (English trans., A. Sisco, 1956, Chicago).

Rogers, Henry Darwin (1858). *The Geology of Pennsylvania*, 2 vols., Blackwood, Edinburgh; Lippincott, Philadelphia.

Rogers, Mrs. William Barton (1896). *Life and Letters of William Barton Rogers*, 2 vols., Riverside Press, Cambridge (Mass.).

Rome De L'Isle, Jean-Baptiste Louis (1772). *Essai de cristallographie . . .* , Paris.

Romer, Alfred (1970). *Henri Becquerel, Dictionary of Scientific Biography,* Vol. 1, Scribners, N.Y.

Rudwick, Martin (1970). The Strategy of Lyell's "Principles of Geology," *Isis,* Vol. 61, pp. 4–33.

Rudwick, Martin (1975). *Charles Lyell, Dictionary of Scientific Biography,* Vol. 12, Scribners, N.Y.

Rudwick, Martin (1979). *The Devonian, a System Born from Conflict, The Devonian System,* M. House et al. eds., Spec. Paper Palaeontology, no. 23, pp. 9–21.

Rülein von Calw, Ulrich (1500). *Eyn nützlich Bergbüchlein,* n. p.

Rutherford, Ernest (1904). *Radio-Activity,* Cambridge Univ. Press.

Rutherford, Ernest and F. Soddy (1902). The Cause and Nature of Radioactivity, pt. 2, ser. 6, Vol. 4, *Phil. Mag.,* pp. 370–396, 569–585.

Rutherford, Ernest and F. Soddy (1902). The Radioactivity of Thorium Compounds, *Trans. of the Chemical Soc.,* Vol. 81, pp. 321–350, 837–860.

Rutot, A. (1913). Découverte d'un nouveau mineur néolithique, à Obourg, *Bull. Soc. Belge Géol., Paleont., et Hydrol.,* Vol. 27, pp. 131–136.

Sachan, E. C. (1964). *Al-Beruni's India,* popular edition, New Delhi.

Savonarola, Girolamo (1534). *Compendivm totivs philosophiae, tam naturalis, & moralis,* Venice.

Scamuzzi, Ernesto (1965). *Egyptian Art in the Egyptian Museum of Turin,* pl. 88, Abrams, New York.

Scherz, Gustav, Ed. (1971). *Dissertations on Steno as Geologist,* University Press, Odense (Denmark).

Scheuchzer, Johann Jacob (1708). *Piscium Querelae et Vindiciae,* Gessner, Zurich.

Scheuchzer, Johann Jacob (1716). *Helvetiae stoicheiographia, orographica et oreographia,* Zurich.

Scheuchzer, Johann Jacob (1726*a*). "Ex hominum diluvio submersorum . . . , letter to Sir Hans Sloane, *Phil. Trans.,* Vol. 34, no. 392, pp. 38–39.

Scheuchzer (1726*b*). *Homo Diluvii Testis,* Burkli, Zurich.

Schmeckebier, Lawrence F. (1904). Catalogue and Index of the Hayden, King, Powell, and Wheeler Surveys, *U.S. Geological Survey Bull.* 222.

Schneer, Cecil J., Ed. (1969). *Toward a History of Geology,* M.I.T. Press, Cambridge (Mass.).

Schneer, Cecil J., Ed. (1979) *Two Hundred Years of Geology in America,* Univ. Press of New England, Hanover.

Schoenflies, Arthur Moritz (1891). *Krystallsysteme und Krystallstructure,* Teubner, Leipzig.

Schoolcraft, Henry R. (1834). *Narrative of an Expedition Through the Upper Mississippi to Itasca Lake . . . ,* Contains D. Houghton, *Report on the Copper of Lake Superior,* Harper, New York.

Schöpf, Johann David (1787*a*). *Beyträge zur mineralogischen Kenntniss des östlichen Theils von Nord Amerika und seiner Gebürge,* Erlangen.

Schöpf, Johann David (1787*b*). *American Materia Medica, Chiefly of the Vegetable Kingdom,* A. Patze, trans., Erlangen.

Schöpf, Johann David (1788). *Reise durch einige der mittlern und südlichen Vereinigten Nordamerikanischen Staaten,* Erlangen.

Schöpf, Johann David (1911). *Travels in the Confederation (1783–1784)*, A. J. Morrison, trans. and ed., 2 vols., Campbell, Philadelphia.

Schöpf, Johann David (1972). *Geology of Eastern North America*, E. M. Spieker, trans. of Schopf, 1787, Hafner, New York.

Schvarcz, Julius (1868). *The Failure of Geological Attempts Made by the Greeks from the Earliest Ages Down to the Epoch of Alexander*, revised & enlarged, Trubner, London.

Schwarzbach, Martin (1980). *Alfred Wegener und die Drift der Kontinente*, Wissenschaftliche Verlagsgesellschaft MBH, Stuttgart.

Scilla, Agostino (1670). *La vana specvlazione disingannata dal senso . . .*, Colicchia, Naples.

Scott, J. F. (1952). *The Scientific Work of René Descartes (1596–1650)*, Taylor & Francis, London.

Scrope, George Poulett (1825). *Considerations on Volcanos*, Phillips and Yard, London.

Scrope, George Poulett (1827). *Maps and Plates to the Memoir on the Geology and Volcanic Formations of Central France*, Longman et al., London.

Scudder, Samuel (1890). The Tertiary Insects of North America, *U.S. Geol. and Geog. Survey of the Territories (Hayden Survey)*, Vol. XII, Government Printing Office, Washington.

Sedgwick, Adam and Roderick Impey Murchison (1840). On the Physical Structures of Devonshire . . . , *Trans. of the Geol. Soc. of London*, 2nd ser., Vol. 5, pt. 3.

Sedgwick, Adam and Roderick Impey Murchison (1842). On the Distribution and Classification of the Older or Palaeozoic Deposits of the North of Germany . . . , *Trans. of the Royal Soc. of London*, 2nd ser., Vol. 6 (Taylor, London, 1842).

Smith, Edgar Fahs (1919). *Chemistry in Old Philadelphia*, Lippincott, Philadelphia.

Smith, I. F. (1974). The Neolithic, in C. Renfrew, ed., *British Prehistory: A New Outline*, Duckworth, London, pp. 100–136.

Snider-Pellegrini, Antonio (1858). *La création et ses mystères dévoilés*, Franck & Dentu, Paris.

Sorby, Henry Clifton (1853). On the Origin of Slaty-Cleavage, *Edinburgh New Phil. Jour.*, Vol. 55, pp. 137–150.

Sorby, Henry Clifton (1858). On the Microscopical Structure of Crystals, Indicating the Origin of Minerals and Rocks, *Jour. Geol. Soc. London*, Vol. 14, pp. 453–500.

Sorby, Henry Clifton (1870). On the Application of the Microscope to the Study of Rocks, *Monthly Micro. Jour.*, Vol. 4., pp. 148–149.

Soulavie, Jean-Louis Giraud (1780–1784). *Histoire naturelle de la France Méridionale*, 8 vols., Paris.

Sowerby, James (1804–1817). *British Mineralogy . . .*, 5 vols., London.

Sowerby, James (1811). *Exotic Mineralogy . . . Supplement to British Mineralogy*, pp. 1–145, London.

Sowerby, James (1812). *The Mineral Conchology of Great Britain*, Vol. 1, no. 12, London.

Sowerby, James and James de Carle Sowerby (1812–1846). *The Mineral Conchology of Great Britain*, 7 vols., London.

Stanton, William R. (1975). *The Great United States Exploring Expedition of 1838–1842*, Univ. of California Press, Berkeley.

Steno, Nicolaus (1667). Elementorvm myologiae specimen sev musculi descriptio geometrica. *Cvi Accedvnt Canis carchariae dissectvm Capvt, et Dissectvs piscis ex canvm genere*, Florence.

Steno, Nicolaus (1669). *De solido intra solidvm naturaliter contento dissertationis prodromvs*, Florence.

Steno, Nicolaus (1916). *The Prodromus of Nicolaus Steno's Dissertation Concerning a Solid Body Enclosed by Process of Nature Within a Solid*, John Garrett Winter, trans., Univ. Michigan Studies, Humanistic Series, Vol. 9, part 2; reprinted Hafner, N.Y. (1968).

Steno, Nicolaus (1969). *Steno Geological Papers*, Gustav Scherz, ed., Alex J. Pollock, trans., Univ. Press, Odense (Denmark).

Stille, Hans (1924). *Grundfragen der vergleichenden Tektonik*, Berlin.

Strachey, John (1719). A Curious Description of the Strata Observed in the Coal Mines of Mendip in Somersetshire, *Phil. Trans. Royal Soc.*, Vol. 30, pp. 968–973.

Strutt, Robert J. (1905). On the Radioactive Minerals, *Royal Soc. London, Proc.*, ser. A, Vol. 76, pp. 83–101.

Suess, Eduard (1875). *Die Entstehung der Alpen*, Braumüller, Vienna.

Suess, Eduard (1883–1909). *Das Antlitz der Erde*, 3 vols., Tempsky, Prague; Vienna, Leipzig.

Suess, Eduard (1897–1918). *La Face de la Terre*, 4 vols., Colin, Paris.

Suess, Eduard (1904–1924). *The Face of the Earth*, 5 vols., Clarendon Press, Oxford.

Sullivan, Robert (1971). *The Disappearance of Dr. Parkman*, Little, Brown, & Co., Boston.

Sweet, Jessie M. (1969). Robert Jameson and Shetland: a Family History, *Scottish Geneologist*, Vol. 16., no. 1, pp. 1–18.

Sweet, Jessie M. (1972). Instructions to Collectors: John Walker (1793) and Robert Jameson (1817) . . . , *Ann. Science*, Vol. 29, no. 4, pp. 397–414.

Sweet, Jessie M. and C. D. Waterston (1967). Robert Jameson's Approach to the Wernerian Theory of the Earth, 1796, *Annals of Science*, Vol. 23, no. 2, pp. 81–95.

Taylor, Frank Bursley (1908). A Review of the Great Lakes History . . . , *Science*, n.s., Vol. 27, pp. 725–726.

Taylor, Frank Bursley (1910). Bearing of the Tertiary Mountain Belts on the Origin of the Earth's Plan, *Bull. Geol. Soc. Amer.*, Vol. 21, pp. 179–226.

Theophrastus (1866). *Theophrasti Eresii Opera, Quae Supersunt, Omnia*, Fredericus Wimmer, trans. & ed., Paris.

Theophrastus (1956). *Theophrastus on Stones*, Earle A. Caley and John F. C. Richards, trans. & ed., Ohio State Univ., Columbus.

Thomson, Helen (1971). *Murder at Harvard*, Houghton Mifflin, Boston.

Thomson, William (1862*a*). On the Age of the Sun's Heat, *Macmillan's Magazine*, Vol. 5, pp. 288–293.

Thomson, William (1862*b*). On the Secular Cooling of the Earth, *Royal Soc. of Edinburgh Trans.*, Vol. 23, no. 1, pp. 157–169. Reprinted in Phil. Mag., Vol. 25, pp. 1–14.

du Toit, Alexander (1937). *Our Wandering Continents*, Oliver & Boyd Ltd., Edinburgh.

Topham, Edward (1776). *Letters from Edinburgh, Written in the Years 1774 and 1775, Dodsley, London.*

Totten, Stanley M. (1980). Frank B. Taylor's Personal Claim as Originator of the Continental Drift Theory, *Geol. Soc. America Abstracts with Programs*, Vol. 12, pp. 536–537.

Trask, John Boardman (1853–1856). *Report on the Geology of the Coast Mountains and Part of the Sierra Nevada . . .* , Sacramento.

Trask, John Boardman (1856). *Report on the Geology of Northern and Southern California*, Sacramento.

Tyler, David B. (1968). The Wilkes Expedition . . . (1838–1842), *Mem. Amer. Phil. Soc.*, Philadelphia, Vol. 73, 435 p.

Tyson, Philip Thomas (1850). *Information in Relation to the Geology and Topography of California*, U.S. 31st Cong., 1st session, Senate Ex. Doc. 47, Washington.

Vallemont, (1696). *La physique occulte, ou traité de la Baguette Divinatoire* . . . , Boudot, Paris.

Vallisnieri, Antonio (1715). *Lezione accademica intorno all' origine delle fontane* . . . , Ertz, Venice.

Vallisnieri, Antonio (1721). *Dei corpi marini che sui monti si trovano*, Venice.

Vallisnieri, Antonio (1905). Antonio Vallisnieri e i moderni concetti intorno ai viventi, Lorenzo Camerano, ed., *Memorie reale Accademia scienze (Torino)*, ser. 2, Vol. 55, pp. 69–112.

Van der Gracht, W.A.J.M. van Waterschoot et al. (1928). *Theory of Continental Drift: a Symposium on the Origin and Movement of Land Masses Both Inter-continental and Intra-continental, as Proposed by Alfred Wegener*, American Assoc. of Petroleum Geol., Tulsa.

Van Rensselaer, Jeremiah (1825). *Lectures on Geology* . . . , Blis and White, New York.

Vine, Fred J. and D. H. Matthews (1963). Magnetic Anomalies over Oceanic Ridges, *Nature*, Vol. 199, pp. 947–949.

Volney, Constantin-François Chasseboef (1803). *Tableau du climat et du sol des États-Unis d'Amérique*, Courcier, Paris.

Volney, Constantin-François Chasseboef (1804). *View of the Climate and Soil of the United States of America* . . . , Johnson, London.

Volney, Constantin-François Chasseboef (1968). A View of the Soil and Climate of the United States of America, C. B. Brown, trans., *Contributions to the History of Geology*, Vol. 2, Hafner, New York.

Vrooman, Jack Rochford (1970). *René Descartes, A Biography*, Putnam, N.Y.

Wagner, Johann Jacob (1680). *Historia naturalis Helvetiae curiosa*, p. 356.

Walcott, Charles D. (1876). *Preliminary Notice of the Discovery of the Remains of the Natatory and Branchial Appendages of Trilobites* (advanced print), 28th Annual Rep., New York State Museum Natural History (1879), pp. 89–92.

Walcott, Charles D. (1890). The Fauna of the Lower Cambrian or *Olenellus* Zone, *U.S. Geol. Survey Annual Rpt.*, pt. 1, pp. 509–774.

Walcott, Charles D. (1899). Pre-Cambrian Fossiliferous Formations, *Bull. Geol. Soc. America*, Vol. 10, pp. 199–414.

Walker, John (1966). *Lectures on Geology by John Walker*, H. Scott, ed. Univ. of Chicago Press.

Warren, Erasmus (1690). *Geologia: or a Discourse Concerning the Earth before the Deluge*, Southby, London.

Warren, Erasmus (1691). *Defence of His Exceptions (in controversy with Burnet)*, see Porter (1977), pp. 83–84.

Waller, Richard, ed. (1705). *The Posthumous Works of Robert Hooke*, Smith & Walford, London.

Watson, Sereno (1871). Botany, *U.S. Geol. Exploration of the Fortieth Parallel (King Survey)*, Vol. V, Government Printing Office, Washington.

Watson, White (1811). *A Delineation of the Strata of Derbyshire . . .*, Todd, Sheffield.

Watt, Gregory (1804). Observations on Basalt . . . , *Phil. Trans. Royal Soc. London*, Vol. 94, pp. 277–314.

Webster, John White (1821). *Description of the Island of St. Miguel . . .*, Williams, Boston.

Wegener, Alfred (1912). Die Entstehung der Kontinente, *Petermanns Geographische Mitteilungen*, Vol. 58, pp. 185–195, 253–256, 305–309.

Wegener, Alfred (1915). *Die Entstehung der Kontinente und Ozeane*, Friedrich Vieweg & Sohn, Braunschweig.

Wegener, Alfred (1920). *Die Entstehung der Kontinente and Ozeane*, revised ed., Friedrich Vieweg & Sohn, Braunschweig.

Wegener, Alfred (1922). *Die Entstehung der Kontinente and Ozeane*, revised ed., Friedrich Vieweg & Sohn, Braunschweig.

Wegener, Alfred (1924). *The Origin of Continents and Oceans*, Methuen, London.

Wegener, Alfred (1929). *Die Entstehung der Kontinente und Ozeane*, 4th ed., revised, Vieweg & Sohn, Braunschweig.

Wells, John (1959). Notes on the Earliest Geological Maps of the United States 1756–1832, *Jour. Washington Acad. Sci.*, Vol. 49, pp. 198–205.

Wells, John (1963). *Early Investigations of the Devonian System in New York, 1656–1836*, Geol. Soc. America Special Paper 74.

Werner, Abraham Gottlob (1774). *On the External Characters of Minerals*, Albert Carozzi, trans. (1962), Univ. of Chicago Press, Chicago.

Werner, Abraham Gottlob (1786). *Short Classification and Description of the Various Rocks*, A. Ospovat, trans. (1971), Hafner, N.Y.

Werner, Abraham Gottlob (1791*a*). *Ausführliches und sistematisches Verzeichnis des Mineralkabinets . . .*, Freiberg.

Werner, Abraham Gottlob (1791*b*). *Neue Theorie von der Entstehung der Gänge . . .*, Freiberg.

Wheeler, George M. (1889). *Report upon United States Geographical Surveys West of the One Hundredth Meridian*, Vol. 1, Geographical Rpt., Government Printing Office, Washington.

Whiston, William (1696). *A New Theory of the Earth . . .*, Tooke, London.

White, George (1977). William Maclure's Maps of the Geology of the United States, *Soc. Bibliography Nat. History*, Vol. 8, pp. 266–269.

Whitehurst, John (1778). *An Inquiry into the Original State and Formation of the Earth . . .*, London.

Whitney, Josiah Dwight (1854). *The Metallic Wealth of the United States*, Lippincott, Philadelphia.

Whitney, Josiah Dwight (1880). *The Auriferous Gravels of the Sierra Nevada of California*, Harvard Univ. Press, Cambridge (Mass.).

Wiechert, Emil (1908). *Our Present Knowledge of the Earth*, Rpt. of the Board of Regents of the Smithsonian Institution.

Wilkes, John Charles (1844). *Narrative of the United States Exploring Expedition During the Years 1838–1842*, 5 vols., atlas, Philadelphia.

Wilkes, John Charles (1978). *Autobiography of Rear Admiral Charles Wilkes, U.S. Navy*, W. J. Morgan et al., eds., Dept. of the Navy, Washington.

Wilkins, Thurman (1958). *Clarence King, a Biography*, Macmillan, New York.

Wilson, Leonard (1972). *Charles Lyell, the Years to 1841: the Revolution in Geology*, Yale Univ. Press, New Haven.

Winchell, Alexander (1870). *Sketches of Creation . . .*, Harper, New York.

Winchell, Alexander (1874). *The Doctrine of Evolution . . .*, Harper, New York.

Winchell, Alexander (1877). *Reconciliation of Science and Religion*, Harper, New York.

Winchell, Alexander (1880). *Preadamites*, Griggs, Chicago.

Winchell, Alexander (1881). *Sparks from a Geologist's Hammer*, Griggs, Chicago.

Winchell, Alexander (1883). *World Life or Comparative Geology*, Griggs, Chicago.

Winchell, Alexander (1886). *Geological Studies, or Elements of Geology*, Griggs, Chicago.

Winthrop, John (1755). *A Lecture on Earthquakes*, Edes and Gill, Boston.

Winthrop, John (1757–1758). An Account of the Earthquake Felt in New England, and the Neighboring Parts of America, on the 18th of November 1755, *Phil. Trans. of the Royal Soc.*

Wollaston, W. H. (1809). The Description of a Reflective Goniometer, *Phil. Trans. of the Royal Soc. of London*, pp. 253–258.

Woodward, Horace B. (1908). *The History of the Geological Society of London*, Longmans, Green, and Co., London.

Woodward, John (1695). *An Essay Toward a Natural History of the Earth . . .*, Wilkin, London.

Woodward, John (1696). *Brief Instructions for Making Observations in all Parts of the World*, Wilkin, London.

Woodward, John (1729). *An Attempt Towards a Natural History of the Fossils of England; in a Catalogue of the English Fossils in the Collection of J. Woodward, M.D. . . .*, Vol. I, Fayram, Senex, & Osborn & Longman, London.

Zirkel, Ferdinand (1876). Microscopical Petrography, *U.S. Geol. Expl. of the 40th Parallel (King Survey)*, Vol. VI, Government Printing Office, Washington.

Index